Strain Hardening Cement Composites: Structural Design and Performance

RILEM STATE-OF-THE-ART REPORTS

Volume 6

RILEM, The International Union of Laboratories and Experts in Construction Materials, Systems and Structures, founded in 1947, is a non-governmental scientific association whose goal is to contribute to progress in the construction sciences, techniques and industries, essentially by means of the communication it fosters between research and practice. RILEM's focus is on construction materials and their use in building and civil engineering structures, covering all phases of the building process from manufacture to use and recycling of materials. More information on RILEM and its previous publications can be found on www.RILEM.net.

The RILEM State-of-the-Art Reports (STAR) are produced by the Technical Committees. They represent one of the most important outputs that RILEM generates – high level scientific and engineering reports that provide cutting edge knowledge in a given field. The work of the TCs is one of RILEM's key functions.

Members of a TC are experts in their field and give their time freely to share their expertise. As a result, the broader scientific community benefits greatly from RILEM's activities.

RILEM's stated objective is to disseminate this information as widely as possible to the scientific community. RILEM therefore considers the STAR reports of its TCs as of highest importance, and encourages their publication whenever possible.

The information in this and similar reports is mostly pre-normative in the sense that it provides the underlying scientific fundamentals on which standards and codes of practice are based. Without such a solid scientific basis, construction practice will be less than efficient or economical.

It is RILEM's hope that this information will be of wide use to the scientific community.

For further volumes:
http://www.springer.com/series/8780

Keitetsu Rokugo • Tetsushi Kanda
Editors

Strain Hardening Cement Composites: Structural Design and Performance

State-of-the-Art Report of the RILEM Technical Committee 208-HFC, SC3

Springer

Editors
Keitetsu Rokugo
Gifu University
Gifu
Japan

Tetsushi Kanda
Kajima Technical
Research Institute
Tokyo
Japan

ISBN 978-94-024-0607-8 ISBN 978-94-007-4836-1 (eBook)
DOI 10.1007/978-94-007-4836-1
Springer Dordrecht Heidelberg New York London

Softcover re-print of the Hardcover 1st edition 2013

Printed on acid-free paper

Springer is part of Springer Science+Business Media (www.springer.com)

Contents

Chapter 1
Introduction

Strain Hardening Cement Composites, SHCC hereafter, demonstrate excellent mechanical behavior showing tensile strain hardening and multiple fine cracks. This strain hardening behavior improves the durability of concrete structures employing SHCC and the multiple fine cracks enhance structural performance. Reliable tensile performance of SHCC enables us to design structures explicitly accounting for SHCC's tensile properties. Reinforced SHCC elements (R/SHCC) indicate large energy absorbing performance under large seismic excitation.

Reinforced concrete structures are subjected to various types of loading, such as dead load, live load, and seismic load. Against those loads, R/SHCC elements can be designed by superimposing re-bar performance and SHCC's tensile performance. The estimation of SHCC's shear resistance contribution in design, which is deduced from SHCC's tensile property, requires various investigations since the shear resistance mechanism is usually complicated.

This State-of-the-Art Report is written collectively by members of Sub-committee three of the RILEM TC 208-HFC. The report focuses on flexural design, shear design, FE modeling and anti-seismic design of R/SHCC elements as well as application examples. Establishing design methods for new materials usually leads to exploring application areas and this trend should be demonstrated by collecting actual application examples of SHCC in structures.

Exploring SHCC applications necessitates establishing structural design and construction practice manuals at a very early opportunity. Hence, the Japan Society of Civil Engineers published the "Recommendations for design and construction of high performance fiber reinforced cement composite with multiple cracks" in 2007, which contains technical knowledge accumulated in conjunction with the activity of RILEM TC 208-HFC. This recommendation can be freely obtained from the website of JSCE. The RILEM TC 208-HFC Subcommittee three activities described in this report are intended to develop more reliable and practical recommendations by reflecting the latest research results.

(K. Rokugo)

K. Rokugo and T. Kanda (eds.), *Strain Hardening Cement Composites: Structural Design and Performance*, RILEM State-of-the-Art Reports 6,
DOI 10.1007/978-94-007-4836-1_1, © RILEM 2013

Chapter 2
Structural Behavior and Design of R/SHCC Elements

Abstract Structural behavior of steel reinforced SHCC (R/SHCC) elements are discussed in this chapter. Based on the discussion, design approaches for R/SHCC are also investigated. First, in flexure, R/SHCC appears to show similar behavior to regular R/C and its design can be achieved in extension of R/C design by considering short fiber contribution against flexural tensile stress. Second, the shear behavior of SHCC clarified by Finite Element analysis is reviewed. Then the state of the art in shear design approach is summarized. Finally, structural ductility under seismic loading and seismic design concept for beams and columns are discussed.

Keywords Flexural capacity • Shear capacity • Cracking • Ductility • Macro model • Finite element analysis

2.1 Flexural Characteristics and Design

2.1.1 Flexural Capacity of SHCC Members

2.1.1.1 General

Due to its capability of bearing tensile forces after cracking by the bridging effect of fibers, the design of SHCC flexural members can be streamlined by directly incorporating this effect. In other words, design taking into account the tensile capacity of SHCC becomes possible by assuming a perfect elasto-plastic constitutive model as shown in Fig. 2.1.

To make the effect of SHCC's tensile properties on the cross-sectional performance of flexural members easy to understand, Fig. 2.2 shows the concept of strain and stress distributions of a flexural member cross-section having SHCC on the tension side in comparison with those of an ordinary reinforced concrete member. When SHCC is used on the tension side, the entire cross-section including SHCC

K. Rokugo and T. Kanda (eds.), *Strain Hardening Cement Composites: Structural Design and Performance*, RILEM State-of-the-Art Reports 6, DOI 10.1007/978-94-007-4836-1_2, © RILEM 2013

Fig. 2.1 Tensile stress–strain curve of SHCC

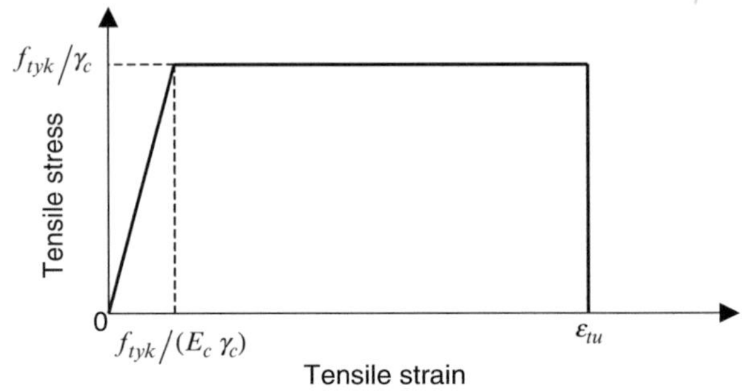

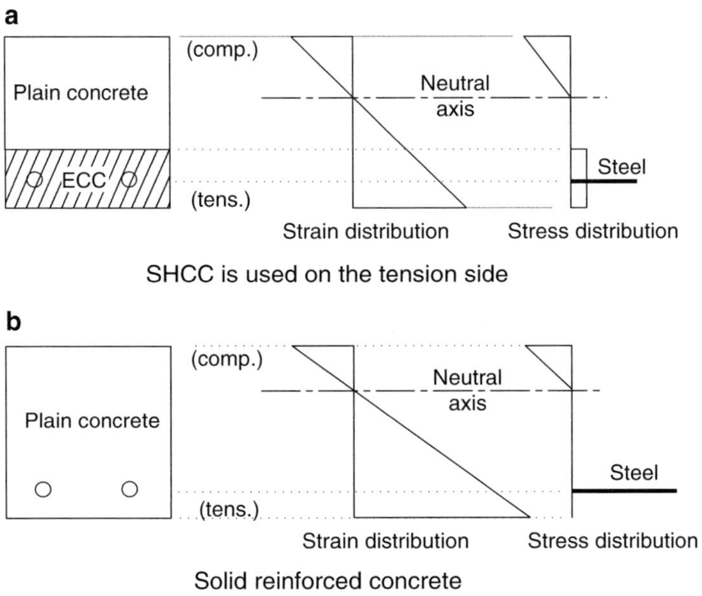

Fig. 2.2 Distribution of stress and strain

bears the tensile forces, reducing the tensile stress of tensile reinforcement even under the action of a tensile strain exceeding the yield strain of reinforcement. Under the same bending moment, the neutral axis of concrete having SHCC on the tension side shifts to the tension side relative to solid reinforced concrete, increasing the compression zone in the concrete. In other words, the use of SHCC on the tension side has the same effect as an increase in the cross-sectional area of tensile reinforcement.

The compressive stress–strain curve should be assumed to calculate the ultimate capacity of section failure for the members subjected to a bending moment or a bending moment and axial compressive forces. An example of a compressive stress–strain relationship, which is derived from the results of tests on cylinder specimens

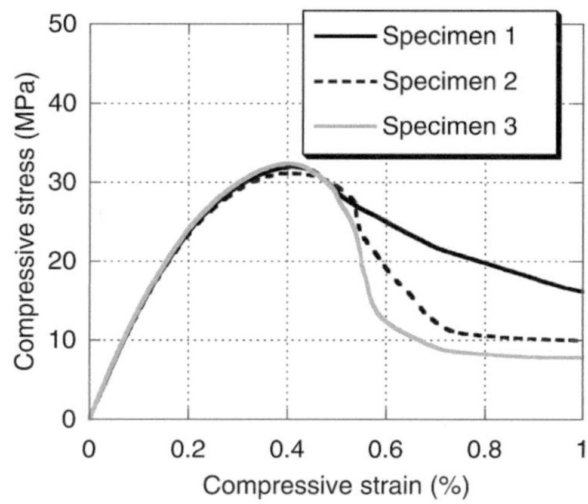

Fig. 2.3 Example of compressive stress–strain curve

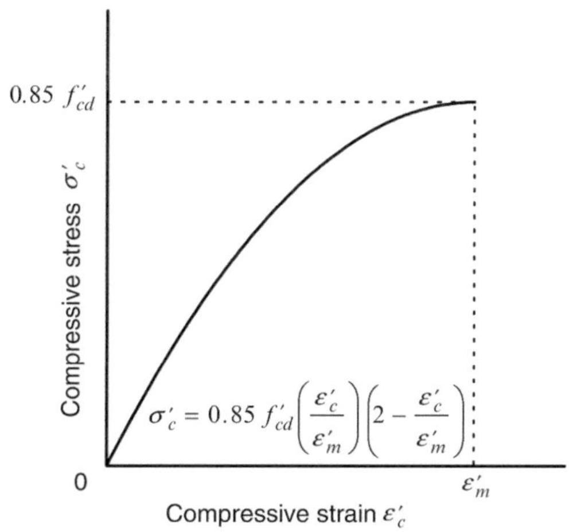

Fig. 2.4 Compressive stress–strain curve in JSCE Recommendations

of 100 mm in diameter and 200 mm in height, is shown in Fig. 2.3. As shown in the figure, the strain value at the peak stress is 0.4% (0.004) approximately. This strain value is greater than 0.002, i.e. that of ordinary concrete. This shows one of the characteristics of SHCC; the stress decreases slowly with increasing strain after the maximum load due to the confinement effect of fiber bridging.

To take into account these SHCC material characteristics in structural design, it is necessary to determine the compressive stress–strain relation of SHCC. According to the JSCE Recommendations, the stress–strain relation shown in Fig. 2.4 may be used for the calculation of the flexural capacity of the section when appropriate test data are not available [1]. The strain ε'_m at the peak stress can be set equal to the strain at reaching the compressive strength obtained from the compression test on the cylindrical specimen.

The influence of the exact shape of the compressive stress–strain curve on the flexural capacity of SHCC members is marginal, if the compressive strength and the

Table 2.1 Details of beam specimens using ECC

Specimen			ECC1	ECC2	RC
Dimensions of specimen	Web width (b_w)	mm	600		
	Effective depth (d)	mm	132		
	Shear span length (a)	mm	600		
	Shear span ratio (a/d)	–	4.6		
	Size of steel bar	Unit	*D*16		
	Area of tensile reinforcement (A_s)	mm^2	1,477		
	Tensile reinforcement ratio ($A_s/(b_w \cdot d)$)	%	1.9		
SHCC	Comp. strength	N/mm^2	45.5	42.6	–
	Ultimate comp. strain	%	0.4	0.4	–
	Young's modulus	kN/mm^2	16.1	15.7	–
	Tensile yield strength	N/mm^2	3.2	2.6	–
	Ultimate tensile strain	%	0.9	1.6	–
Plain concrete	Compressive strength	N/mm^2	38.4	41.6	38.4
	Young's modulus	kN/mm^2	31.0	28.2	31.0
	Tensile strength	N/mm^2	2.7	3.3	2.7

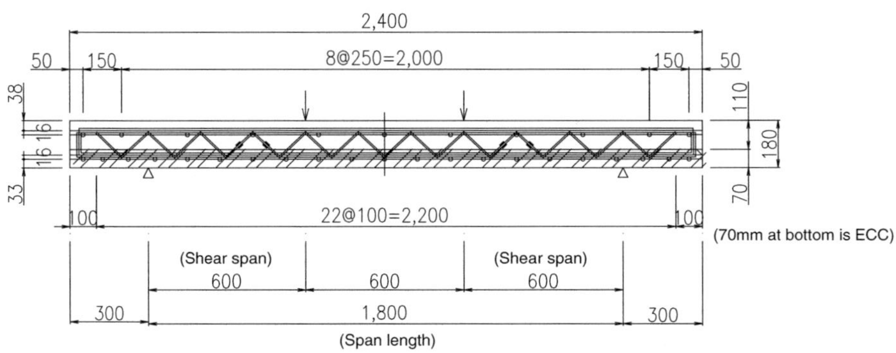

Fig. 2.5 Beam specimen

ultimate strain are the same. However, the ultimate compressive strain and the shape of stress–strain curve especially in the descending region, e.g., shown in Fig. 2.3, are important to estimate the deformation capacity of SHCC members reinforced with tensile steel bars.

2.1.1.2 Bending Tests on Beam Members

Table 2.1 and Fig. 2.5 show the specifications of tests using ECC, as an example of bending testing on beam members having SHCC. Figures 2.6 and 2.7 show the test values compared with the analyzed values by the fiber method(e.g., [2]).

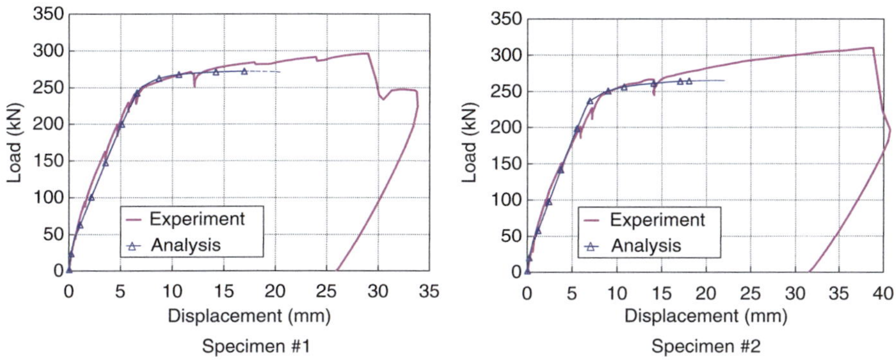

Fig. 2.6 Load–displacement curve, *left*: specimen #1, and *right*: specimen 2

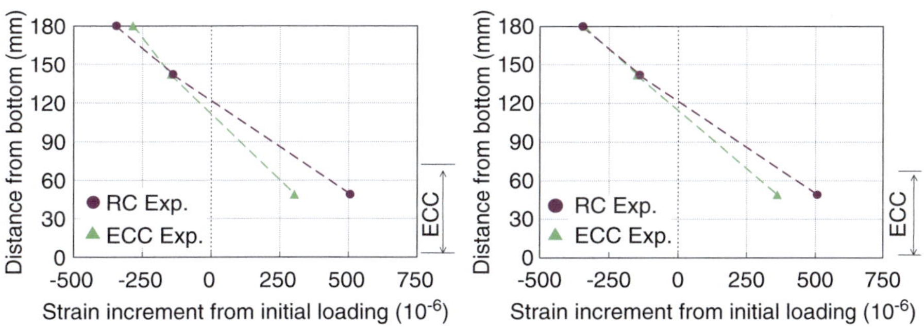

Fig. 2.7 Strain distribution of the section at mid span

Figure 2.6 reveals that the analyzed flexural capacity of beams having ECC is around 90% of the test values similar to the analysis of normal reinforced concrete, proving that the flexural capacity of flexural members having ECC on the tension side can be accurately estimated by the fiber method. Also, Fig. 2.7 reveals that the assumption that the cross-section of a plane remains plane holds by unifying the cross-section of a member strengthened with ECC by an adequate method. The stress in the tensile reinforcement in ECC specimens is lower than that of reference reinforced concrete specimens under the same load, confirming that the use of ECC on the tension side of a flexural member has the same flexural strengthening effect as an increase in the cross-sectional area of reinforcement.

2.1.1.3 Comparison with the Simple Calculation Method

When SHCC is applied to the tension side of a reinforced concrete structure, on which no axial forces act, to bear the load with its entire cross-section, the flexural capacity can be calculated by Eqs. (2.1) and (2.2), assuming the distributions of

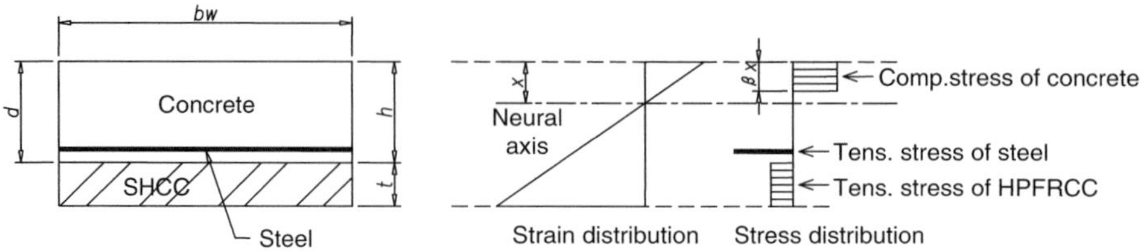

Fig. 2.8 Equivalent stress block, *left*: section of a beam, and *right*: stain distribution

Table 2.2 Comparison of ultimate load carrying capacity

Specimen	Experiment (kN·m)	Analysis (kN·m)	Exp./Ana.
ECC-1	88.8	86.8	1.02
ECC-2	93.0	82.2	1.13

the compressive stress in concrete and the tensile stress in SHCC to be rectangular distributions of compressive stress (equivalent stress blocks) shown in Fig. 2.8. It is necessary, however, that cohesion between SHCC and concrete be adequately maintained. Table 2.2 compares the results of the simple calculation method and the test values. Since the calculations assess the test values on the safe side, this simple calculation method using equivalent stress blocks can be applied in the case where SHCC is placed on the tension side of a reinforced concrete structure under no axial forces and bears the tension with its entire cross section. However, it is necessary to confirm that the strain at the extreme compression fiber reaches the ultimate strain (0.0035) before the tensile strain of SHCC exceeds its ultimate value.

$$x = \frac{p_w d\, f_{yd} + t\, f_{ty}}{k_1 \beta f'_{cd}} \tag{2.1}$$

$$M_u = k_1\beta\left(1 - \frac{\beta}{2}\right) b_w f'_{cd} x^2 + A_s f_{yd}\,(d - x) + f_{ty} b_w t \left(h - x + \frac{t}{2}\right) \tag{2.2}$$

where

p_w:$A_s/\left(b_w \cdot d\right)$
A_s: Cross-sectional area of steel on the tension side
f_{yd}: design yield strength of steel on the tension side
f_{ty}: design tensile yield strength of SHCC
k_1: strength reduction coefficient, generally assumed to be 0.85
$\beta = 0.52 + 80\varepsilon'_{cu}$

$$\varepsilon'_{cu} = \frac{155 - f'_{ck}}{30000} \le 0.0035$$

f'_{ck}: design compressive strength of concrete, but $f'_{ck} \le 80\ \text{N/mm}^2$

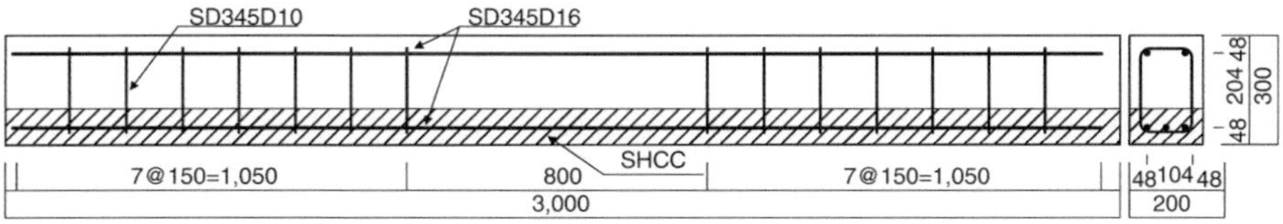

Fig. 2.9 Test specimen for evaluation of flexural cracking of SHCC

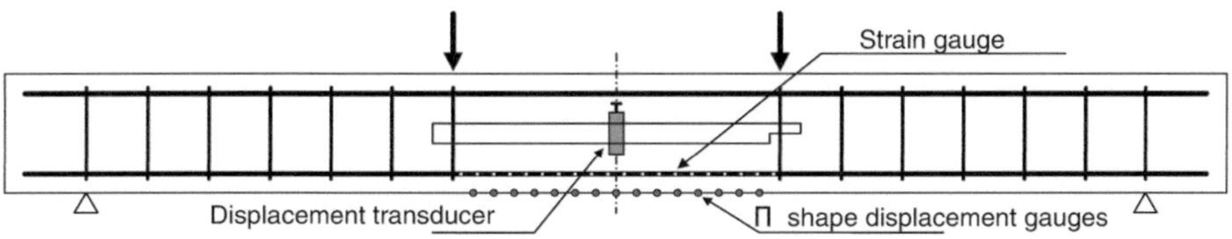

Fig. 2.10 Loading and measurement method

2.1.2 Flexural Cracking Behavior

2.1.2.1 General

A SHCC is characterized by its "strain-hardening property," with which it increases its tensile load (or tensile stress) as its deformation (or strain) increases without causing brittle failure under tensile forces, and "multiple cracking property," with which it disperses cracks into multiple fine ones. The cracking behavior of reinforced concrete beams patched with SHCC was experimentally investigated [3].

2.1.2.2 Outline of the Experiments

Figure 2.9 shows the shape and dimensions of the beam specimen. It is a double-reinforced concrete beam measuring 200 by 300 by 3,000 mm. Assuming bending failure, two and three deformed D16 bars (SD345) were arranged on the compression and tension sides, respectively. The joint surface of specimens to be patched with SHCC was treated using a water jet. Figure 2.10 shows the flexural loading test for beam specimens. The flexural loading tests consisted of four-point loading tests with a moment span of 800 mm. The crack width on the tension side was measured using 15 π shape displacement gauges with a gauge length of 50 mm continuously attached to the bending moment zone. The reference reinforced concrete beam with same dimension was also examined to compare with SHCC patched beam.

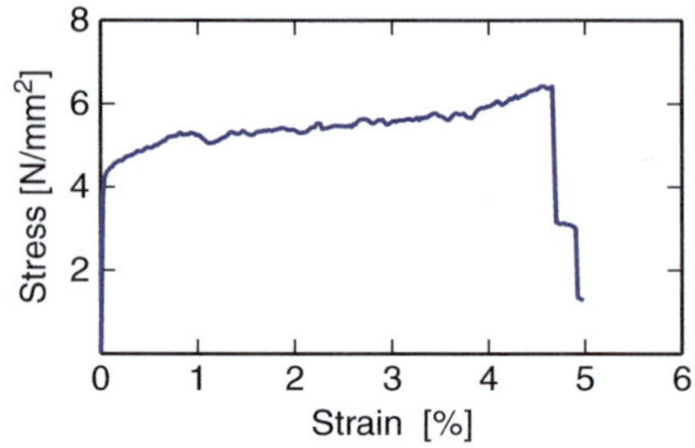

Fig. 2.11 Uniaxial stress–strain curve of SHCC

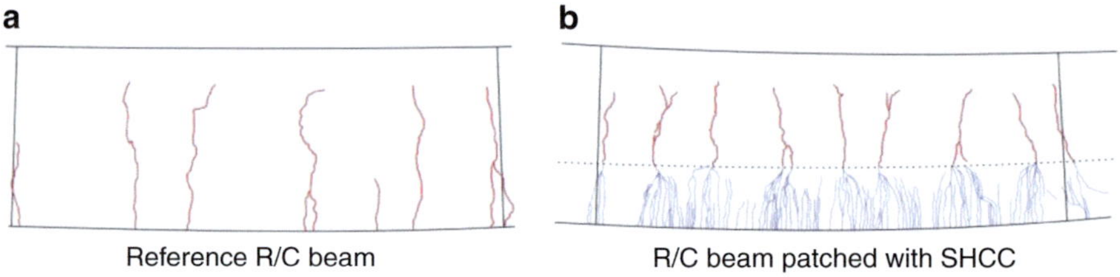

Fig. 2.12 Crack pattern

2.1.2.3 Test Results

Figure 2.11 shows the uniaxial tension test results of SHCC. The ultimate strain and tensile strength of SHCC were 4.7% and 6.4 kN/mm^2, respectively.

The state of cracking during the loading test in reference R/C specimens significantly differed from that in specimens patched with SHCC. Cracks occurred on the tension edge of reference RC specimens at 15–20 cm intervals, with their width increasing as the load increased. On the other hand, dispersed fine cracks occurred on the tension edge of specimens patched with SHCC as the load increased. When cracking occurred in the concrete substrate, the cracks in SHCC tended to concentrate on the locations of cracks in substrate concrete. However, the number of fine cracks increased meanwhile on the tension edge, without crack concentration on the tension edge (see Figs. 2.12 and 2.13). Note that the crack intervals in the concrete substrate were around 10 cm, being evidently narrower than in the reference R/C specimens. Also, no crack was visually observed along the joint between concrete and SHCC until yielding of reinforcement.

Figure 2.14a shows the relationship between the width of a representative crack in the tension edge of a reference R/C specimen measured by a π shape displacement gauge and the strain of reinforcement at the location of the crack. In this case, the crack width increased in proportion to the reinforcement strain.

Figure 2.14b shows the relationship between the width of a representative crack measured in a specimen patched with SHCC and the reinforcement strain. In the figure, the solid line represents the crack width measured by a π shape displacement gauge, increasing as the reinforcement strain increases. However, the slope of the curve is evidently gentler than the case of reference specimen shown in Fig. 2.14a.

Fig. 2.13 Fine cracks in SHCC

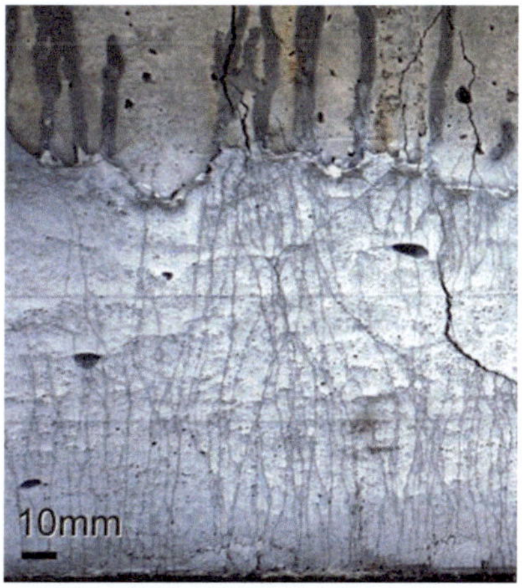

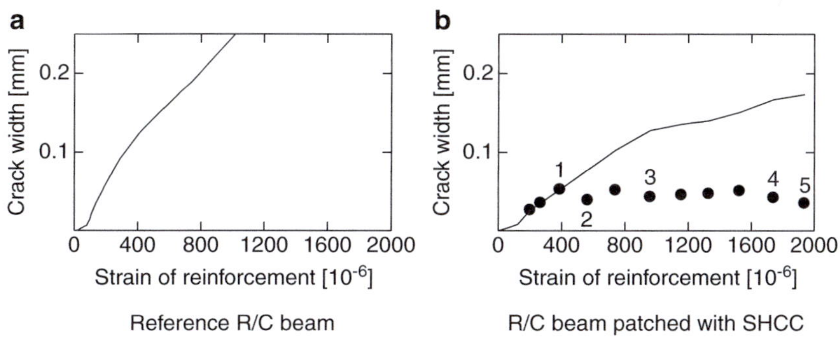

Fig. 2.14 Crack width and strain in steel reinforcement

For specimens patched with SHCC, loading was stopped at certain increments in the load to visually count the number of cracks present within the measurement length of each π shape gauge. The black circles in the figure represent the values obtained by dividing the measured displacement by the number of cracks, which correspond to the average crack width within the measurement range of each π shape gauge. Also, the numbers in the figure denote the number of cracks. Accordingly, the average crack width of specimens patched with SHCC is found to increase to around 0.1 mm as the reinforcement strain increases but scarcely increase thereafter by increasing the number of cracks instead.

2.1.2.4 Evaluation of Flexural Crack Width in JSCE Recommendations [1]

In the ordinary RC members, the flexural crack width increases with the tensile stress in the steel reinforcement. On the other hand, SHCC has the crack width

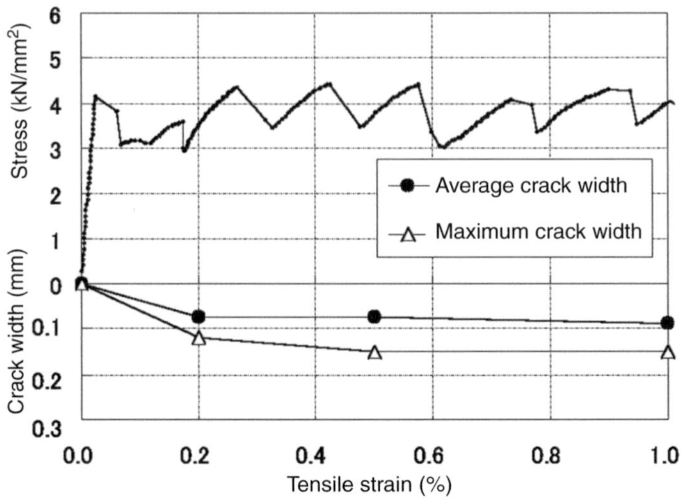

Fig. 2.15 Relation of tensile strain and crack width of SHCC

control capability itself. As mentioned above, the crack width of SHCC reinforced with steel bar increases to a certain width as the rebar strain increases but scarcely increase thereafter by increasing the number of cracks instead. The experimental results showed that the crack width of SHCC reinforced with steel bar is equivalent to or smaller than that without steel reinforcement. Thus, the evaluation of crack width can be made by estimating the strain of SHCC in the members providing that the relation between crack width and tensile strain of SHCC as shown in Fig. 2.15 is determined by uniaxial tensile test in advance. The test method for crack width of SHCC is specified in JSCE Recommendations of HPFCC for the purpose of crack width evaluation in structural members.

2.1.3 *Size Effect on Flexural Behavior*

Kanakubo et al. reported test results concerning the size effect on flexural behavior of SHCC beam specimens [4]. The flexural strength and deformation capacity decrease as the size of beams increase. One of the reasons for this is considered size effect, which is due to the fiber orientation in the specimens. For example, the fiber orientation tends to be two or one dimension at the small size specimens.

Figure 2.16 shows the crack patterns of the SHCC beam specimens with different height (40–280 mm). The cracks occur over the pure bending regions uniformly in small size specimens, while multiple cracks concentrate in certain region in large specimens. While the number of cracks is larger in large size specimens, crack spacing is smaller in small size specimens.

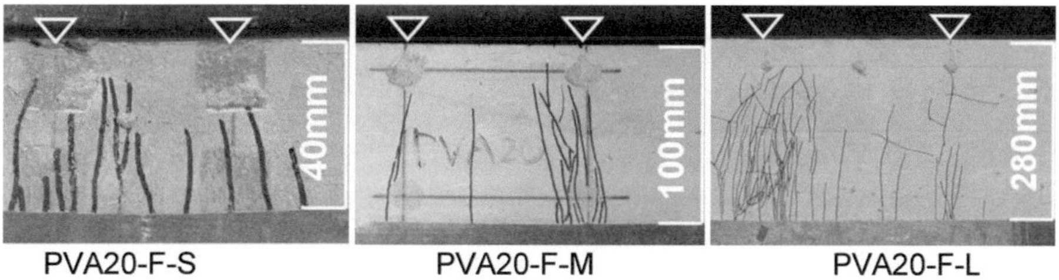

Fig. 2.16 Crack patterns in bending specimens

Fig. 2.17 Relation between high-stress volume and tensile strength

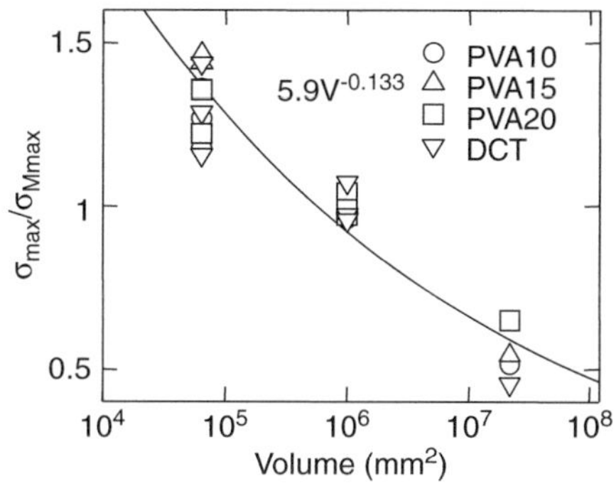

Fig. 2.18 Flexural strength of SHCC beams reinforced with steel bars

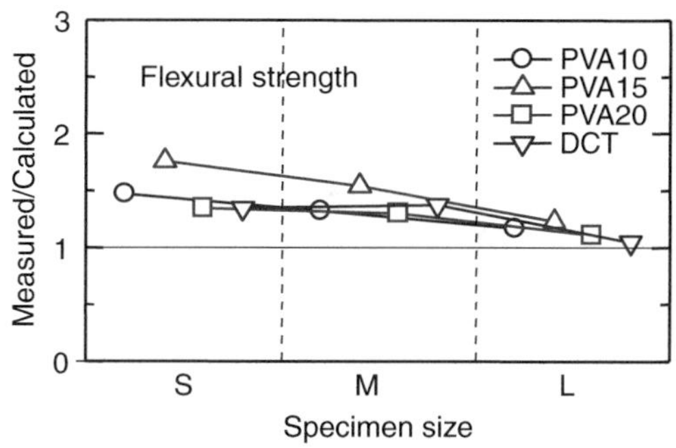

Figure 2.17 shows tensile strength obtained from bending test results on the different size beams. These tensile strengths were estimated according to the method of the Appendix of JCI- S-003-2005 [5]. The horizontal axis represents high-stress volume, which is defined as the volume of pure bending region in beam specimens subjected to two point flexural loading. The tensile strength ($f_{t,b}$) is normalized by that of medium size specimen ($f_{t,b-M}$). The distinct size effect on the tensile strength is shown in Fig. 2.17.

Kanakubo et al. also reported the test and calculated results on the flexural strength of the SHCC beams reinforced with steel bars on the tension side. The flexural strength was calculated by section analysis taking into account of size effect on tensile strength of SHCC shown in Fig. 2.18. The size effect on flexural strength (capacity) was also recognized in SHCC beams reinforced with steel bars.

Contrary to Kanakubo's results, Lepech and Li reported negligible size effects on SHCC's flexural strength [6]. They conducted beam flexural tests with different size specimens with ECC, a representative SHCC type. As a result, flexural strength of the beams did not decrease with beam size up to section height of 200 mm. This contradiction may be led by adopted SHCC's characteristics including tensile strain capacity, and SHCC's size effect is to be further investigated via future research.

(Y. Uchida)

2.2 Shear Characteristics and Design

2.2.1 Introduction

In many civil engineering and building structures, there exist members or details that have to resist intense shear loading, such as coupling beams, shear walls, dampers, beam ends, column bases, etc. When these elements are constructed with reinforced concrete (R/C), they are prone to brittle failure, mainly due to the quasi-brittle character of concrete material. Thus, due to their high tensile strain capacity and damage tolerance, SHCC appears as an attractive alternative for replacing concrete in these critical regions. A large number of experimental studies (see below) have shown that using SHCC in conjunction with conventional reinforcement (reinforced SHCC or R/SHCC) in shear-critical elements indeed leads to significant qualitative improvement of their structural behavior, namely in terms of load and deformation capacity.

The recent transitions from prescriptive to performance-based structural design opens new possibilities for utilizing SHCC materials in engineering practice. However, it is necessary to quantify the performance gains associated with the use of SHCCs and relate them to elementary material properties, such as compressive strength, tensile strength, and tensile strain capacity. Attempts to estimate the shear load carrying capacity of R/SHCC members have been made, mostly along the line of modifying existing formulas for reinforced concrete [8–13]. However, these models have been usually validated on a limited set of experimental data. Presently, there exists no consensus on a universal formula that could reliably predict the shear load carrying capacity of R/SHCC members.

The purpose of the present section is to:

(a) summarize knowledge about shear failure of R/SHCC elements that has been so far gained from experiments by various research groups;
(b) review some of the proposed simplified formulas for prediction of shear load capacity of R/SHCC members and check them against available experimental data;
(c) exemplify the potential of using an FE analysis to clarify a mechanism of shear failure in R/SHCC members.

It is expected that the information presented herein will support formulation of a simplified behavioral model, that could be used in design of R/SHCC members resisting shear loading.

We restrict attention to those cases, when R/SHCC members fail in shear. That is, elements exposed to intense shear but failing in flexure mode, such as reinforced dampers consisting of ECC, a typical material among SHCCs, reported in [9] or precast panels presented in [14] and [15], are not elaborated on here. The main consideration is paid to the cases when shear failure occurs in regions that, prior to cracking, exhibit a smooth stress distribution (beams, panels, etc.). Fracture in locations of shear stress concentration (e.g. stud connections, prestressing strands anchorage [16] etc.) is not discussed.

It should be also noted that only SHCC materials with polymeric fibers (as opposed to those with stiff steel fibers) are discussed in the present chapter.

2.2.2 *Review of Previous Studies*

2.2.2.1 Experiments

Reinforced SHCC Panel under Pure Shear Loading

A unique set of experiments is reported in [11]. Square panels made of reinforced PVA-ECC, which is ECC reinforced with PVA fibers, were tested in a special rig under pure shear loading conditions. Variable parameters were reinforcement ratios and reinforcement yield strengths. All panels showed multiple cracking, evenly distributed throughout the entire specimen. In samples with equal longitudinal and transversal reinforcement ($p_l = p_w$), the cracks were all aligned with the main compressive diagonal. However, in samples with $p_l > p_w$, cracks that formed after yielding of the transversal reinforcement were slightly inclined toward the direction of the heavier longitudinal reinforcement. The specimens with $p_l = p_w$ failed by localized tensile cracks aligned with the compressive diagonal after yielding of reinforcement in both directions. In samples with $p_l > p_w$ failure occurred after yielding of the transversal reinforcement by a sliding crack aligned with the panel edge.

Direct Shear Test under Tension

SHCC has to carry the stresses with the combination of tensile/compressive stress and shear stress along the crack surface in the shear resistant elements such as coupling beams, shear walls, dampers, beam ends, column bases, etc. Though SHCC is well known as the cementitious composites with multiple cracks under tension, the shear behavior along the crack surface remains not fully understood.

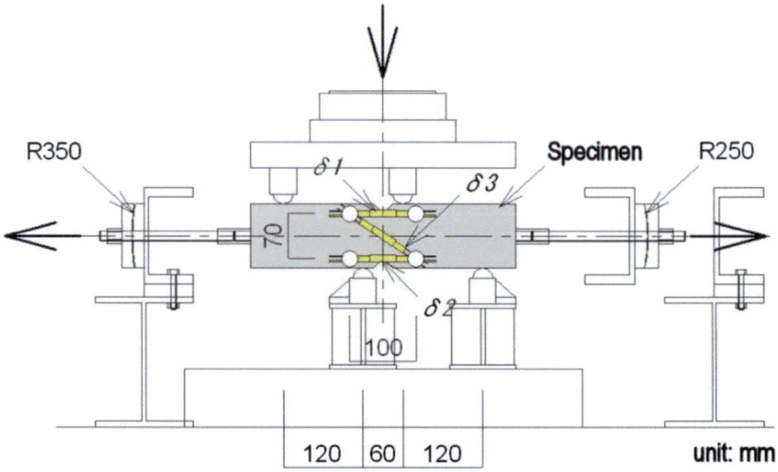

Fig. 2.19 Direct shear test under tension [17]

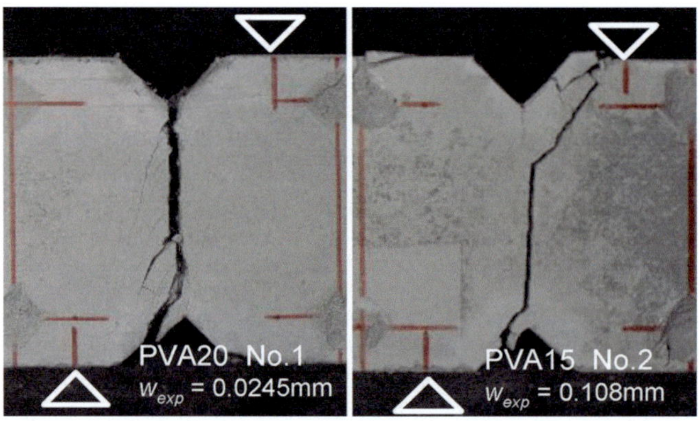

Fig. 2.20 Failure pattern, *left*: ECC with fiber volume fraction 2%, and *right*: ECC with fiber volume fraction 1.5% [17]

This characteristic gives one of the most important information when finite element analysis is employed to investigate the structural behavior of shear resistant elements.

A direct shear test using notched prism specimens is reported in [17]. The loading system is shown in Fig. 2.19. The dimension of specimen is 100 × 100 mm section and 400 mm length. The notches are set on the center of specimen. Direct shear load was applied after the axial tensile loading and first cracking. The axial tensile load was kept constant during the shear loading, so the axial tensile load was almost the same as first cracking strength. Figure 2.20 shows the typical failure pattern. The diagonal cracks were observed after the contacting of both cracked surfaces due to the rotation of the both blocks of specimen.

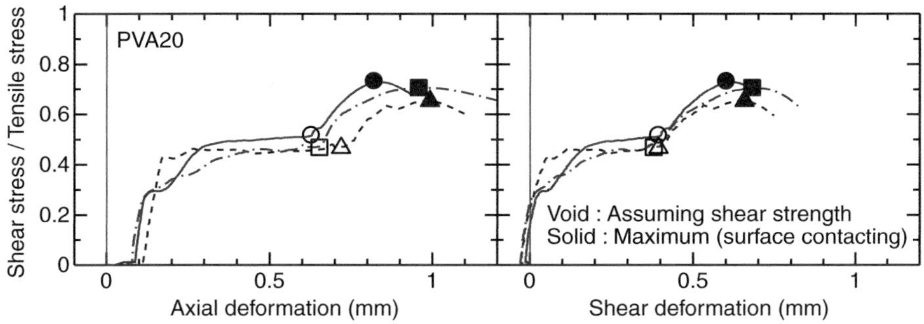

Fig. 2.21 Typical example of shear stress – deformation curve [17]

Figure 2.21 shows the examples of shear stress – deformation curves of the specimens with PVA fiber of $V_f = 2.0\%$. Axial deformation of around 0.1 mm is observed by tensile loading only. After starting shear loading, both axial and shear deformation increase, and the shear stress keep almost the constant value as same as about 0.5 times of tensile stress. After that, the shear stress starts increasing again because of the contacting of cracked surfaces. The report concludes that ECC has the capacity to transmit the tensile stress of the same level of first cracking strength even if the large shear deformation exists.

Bi-directional Multiple Cracking Tests

In order to elucidate the behavior of multiply-cracked SHCC when cracks are exposed to shearing, Suryanto et al. [18, 19] conducted a series of bending tests on pre-cracked PVA-ECC plates. In that study, different levels of multiple cracking were first induced to large ECC plates by subjecting them to 4-point bending (stage 1). Subsequently, smaller plates were produced by cutting the large plates at different angles with respect to the pre-cracked direction. These smaller plates were then loaded in bending (stage 2). The procedure is depicted in Fig. 2.22a. The tests showed that during stage 2 the preexisting cracks were undergoing simultaneous sliding and opening, as shear and tensile stresses developed on multiple cracks. When these stresses were developed, a new set of cracks formed nearly orthogonal to the preexisting ones, forming a bi-directional multiple cracking pattern (see Fig. 2.22b). The orthogonal crack pattern suggests that the preexisting cracks exhibited low shear stress transfer. The initial stiffness, load carrying capacity, and the deflection at the peak load at stage 2 depended on the amount and direction of pre-cracking. The tests showed that as the angle of the precrack direction becomes larger (with respect to the longitudinal axis of the plate), there is an apparent strength reduction, while initial stiffness consistently increases with the precrack direction. Since this macroscopic response is related to the tensile and shear stresses transferred along the bidirectional cracks, the test results can be used to verify modeling of tensile and shear transfer.

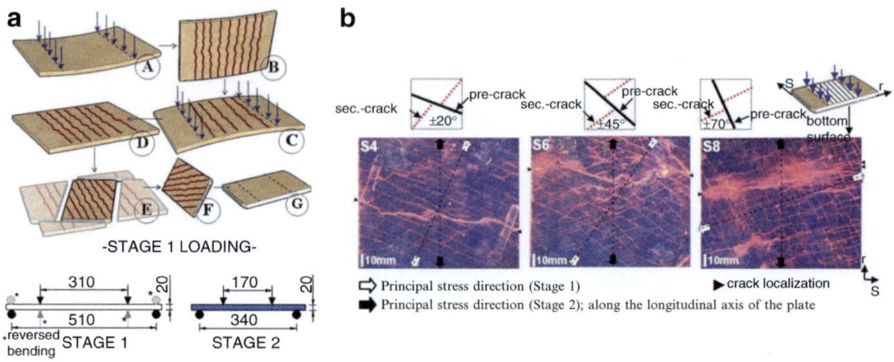

Fig. 2.22 Configuration and procedure of bi-directional multiple cracking tests (a) and typical crack pattern after failure (b) [18]

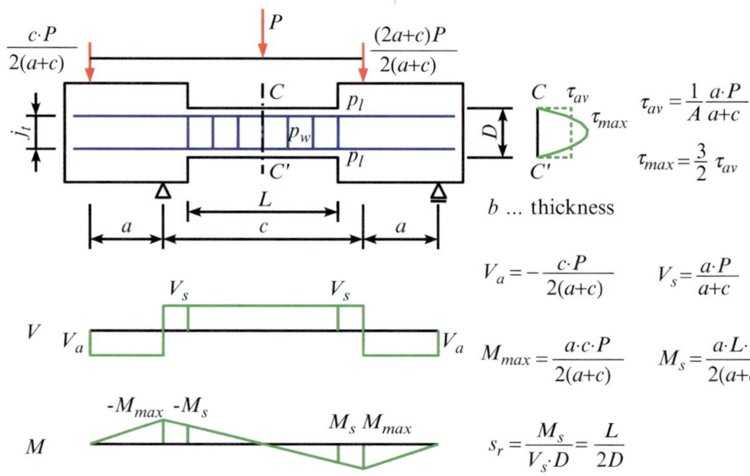

Fig. 2.23 Configuration and parameters of Ohno shear beam

Ohno Shear Beams

In several experimental programs, the so-called Ohno beam was used to assess the shear behavior of R/SHCC. As shown in Fig. 2.23, the beam is configured such that the state of pure shear exists on the central cross-section, if the material is isotropic and homogeneous. The configuration is characterized by shear-span to depth ratio s_r and longitudinal and transversal reinforcement ratios in the shear span, p_l and p_w, respectively (see Fig. 2.23). Material parameters of the experiment are typically yield strength of longitudinal and transversal reinforcement σ_{ly} and σ_{wy}, resp., SHCC strength in compression σ_B, first crack strength σ_{cr}, ultimate strength in tension σ_{max}, and tensile strain capacity ε_u. The beams are loaded either monotonically (M) or by alternating low-cycle shear (A). The behavior of the beam is expressed in terms of average shear stress τ_{av} or shear force V_s at mid-span and shear distortion (deflection angle, rotation angle) γ_{av}. The former two are explained in Fig. 2.23 and the latter is defined as the difference between the vertical displacements at the ends of the shear span, divided by its length L. The experimental studies usually also report the failure mode, which can be flexure (F), bond splitting (BS), or shear (S). Sometimes, shear-tension (ST) and shear-compression (SC) failure modes are distinguished.

The parameters and results of several experimental studies [10, 12, 20–23] are summarized in Table 2.3. In general, all of the experiments demonstrated that, when SHCC was used, the shear beams were undergoing distributed diagonal multiple cracking prior to formation of localized failure-crack(s). This behavior was reflected in improved load and deformation capacity over similarly configured concrete or quasi-brittle FRC specimens. The SHCC beams also exhibited sub-millimeter crack widths and maintained better integrity and resisted cover spalling when exposed to alternating shear load.

Study [21] involved two beams made of ECC with different strength and ductility (SPECC, DRECC) and three control specimens: PC (ordinary concrete without shear reinforcement), RC (ordinary concrete reinforced by wire mesh), and FRC (quasi-brittle steel fiber reinforced concrete). This was the only series among those discussed here, in which the longitudinal reinforcement did not extend over the entire beam, but was terminated in the initially compressed zones of the middle panel. However, tensile stress concentration at the ends of the bars caused initiation of cracks near the beam top and bottom surfaces. These cracks propagated diagonally into the mid-section at the peak load. Therefore the model shown in Fig. 2.23 should not be used to infer the shear strength of the cementitious material and the failure mode is categorized as flexure-shear in Table 2.3.

Variable parameters of experiments described in [12, 22] were beam material (ordinary concrete [RC] and ECC), span ratio s_r (0.5 and 1), and transversal reinforcement ratio (0% and 1%). The ECC beams exhibited improved load carrying capacity over concrete ones under both shear-tension and shear-compression failure modes. Increased deformation capacity was observed when shear-tension failure took place. The tests proved that combining transverse reinforcement with ECC is effective in enhancing shear load capacity, although ECC beams behaved in a ductile manner even without transverse reinforcement. They also demonstrated that using SHCC (even without transversal reinforcement) is effective in preventing brittle shear behavior typical for short-spanned elements.

Control R/C beams in experiments reported in [20] were configured so as to fail in shear (#1) and by bond splitting (#5). Since identically reinforced ECC beams (#2 and #6, resp.) both failed in flexure, the effectiveness of using SHCC to prevent shear and bond splitting failure was demonstrated. Beams #3 and #4 contained high-strength longitudinal reinforcement in order to induce shear failure. Therefore, their results can be used to estimate the shear load carrying capacity of reinforced ECC members.

Table 2.3 Parameters and results of Ohno shear beams

			Experimental parameters and results																		Analytic model results: V_u (exper.)/V_u (model)			
			L	D	b	j_t	s_r	p_l	p_w	σ_{ly}	σ_{wy}	σ_B	σ_{cr}	σ_{max}	σ_{max}[a]	ε_u	V_u	$\tau_{av,u}$	γ_{av} @ peak		Nagai's eq.	Kanakubo's eq.	JSCE's eq.	Suwada's eq.
Ref.	Spec. #	Loading type	m	m	m	m	–	%	%	MPa	MPa	MPa	MPa	MPa	MPa	%	kN	MPa	10^{-2} rad	Reported failure	–	–	–	–
Li et al. [21]	PC	M	0.17	0.21	0.05	0.13	0.40	4.05	0	N/A	0	41.8	2.5	2.5	N/A	0.02	26.04	2.48	0.12	F-S				
	RC	M	0.17	0.21	0.05	0.13	0.40	4.80	0.75	N/A	300[b]	41.8	2.5	2.5	N/A	0.02	57.44	5.47	1.95	F-S				
	FRC	M	0.17	0.21	0.05	0.15	0.40	2.70	0	N/A	0	47.9	4.3	4.3	N/A	0.05	31.82	3.03	0.6	F-S				
	SPECC	M	0.17	0.21	0.05	0.15	0.40	2.70	0	N/A	0	53.5	2.4	4.7	N/A	5.6	53.45	5.09	2.6	F-S				
	DRECC	M	0.17	0.21	0.05	0.13	0.40	4.05	0	N/A	0	89.2	N/A	6.2	N/A	0.15	103.8	9.89	0.7	F-S				
Kanda [12]	RC-1-1	A	0.4	0.2	0.15	0.133	1.00	4	1	563	364	30.5	N/A	2.55	N/A	0.011	129	4.3	0728	SC				
	RC-1-0	A	0.4	0.2	0.15	0.133	1.00	4	0	563	0	33.6	N/A	2.31	N/A	0.009	63	2.1	0.244	ST				
	ECC-1-1	A	0.4	0.2	0.15	0.133	1.00	4	1	563	364	41.2	N/A	3.57	N/A	3.34	186	6.2	0.76	SC	1.28	**0.90**	**0.94**	1.06
	ECC-1-0	A	0.4	0.2	0.15	0.133	1.00	4	0	563	0	45.7	N/A	2.61	N/A	1.4	90	3	0.791	ST	**0.92**	**0.76**	1.00	1.09
	ECC-0.5-1	A	0.2	0.2	0.15	0.133	0.50	4	1	563	364	40.1	N/A	3.38	N/A	1.21	261	8.7	0.66	SC	1.52	1.26	1.36	1.65
	ECC-0.5-0	A	0.2	0.2	0.15	0.133	0.50	4	0	563	0	41.6	N/A	3.38	N/A	1.21	192	6.4	0.595	ST	1.28	1.08	1.79	4.10
Fukuyma et al. [20]	1	A	1.08	0.27	0.2	0.187	2.00	1.41	0.211	393	272	34.9	N/A	N/A	N/A	~0	110	2.037	0.8	ST				
	2	A	1.08	0.27	0.2	0.187	2.00	1.41	0.211	393	272	50.5	1.6	1.8	N/A	1.2	127	2.352	2-3	F				
	3	A	1.08	0.27	0.2	0.187	2.00	1.41	0.211	760	272	50.5	1.6	1.8	N/A	1.2	191	3.537	2	S	1.55	1.14	1.34	1.59
	4	A	1.08	0.27	0.2	0.187	2.00	1.41	0	393	0	50.5	1.6	1.8	N/A	1.2	150	2.778	2-3	F-S				
	5	A	1.08	0.27	0.2	0.216	2.00	1.47	0.423	396	272	34.9	N/A	N/A	N/A	~	141	2.611	0.9	BS				
	6	A	1.08	0.27	0.2	0.216	2.00	1.47	0.423	396	272	50.5	1.6	1.8	N/A	1.2	155	2.87	3	F				

Shimizu et al. [10]	NC-00	M	0.84	0.28	0.18	0.184	1.50	2.43	0	719	294.5	39	N/A	N/A	N/A	~0	116.4	2.31	0.505	S				
	NC-15	M	0.84	0.28	0.18	0.184	1.50	2.43	0.15	719	294.5	39	N/A	N/A	N/A	~0	104.8	2.079	0.693	S				
	NO-30	M	0.84	0.28	0.18	0.184	1.50	2.43	0.3	719	294.5	39	N/A	N/A	N/A	~0	132.3	2.625	1.201	S				
	PVA 10-00	M	0.84	0.28	0.18	0.184	1.50	2.43	0	719.7	294.5	37.3	2.54	2.52	2.62	0.1	123.9	2.458	0.644	S	1.09	**0.86**	**0.90**	1.58
	PVA 10-15	M	0.84	0.28	0.18	0.184	1.50	2.43	0.15	719.7	294.5	37.3	2.54	2.52	2.62	0.1	144.8	2.873	0.807	S	1.16	**0.88**	**0.93**	1.63
	PVA 10-30	M	0.84	0.28	0.18	0.184	1.50	2.43	0.3	719.7	294.5	37.3	2.54	2.52	2.62	0.1	171.5	3.403	1.175	S	1.27	**0.93**	**0.99**	1.27
	PVA 15-00	M	0.84	0.28	0.18	0.184	1.50	2.43	0	719.7	294.5	35.7	2.41	2.5	3.77	1.16	142.8	2.833	0.911	S	1.30	1.01	1.05	1.53
	PVA 15-15	M	0.84	0.28	0.18	0.184	1.50	2.43	0.15	719.7	294.5	35.7	2.41	2.5	3.77	1.16	169.7	3.367	1.101	S	1.41	1.05	1.10	1.44
	PVA 15-30	M	0.84	0.28	0.18	0.184	1.50	2.43	0.3	719.7	294.5	35.7	2.41	2.5	3.77	1.16	182.9	3.629	1.164	S	1.40	1.01	1.07	1.29
	PVA 20-00	M	0.84	0.28	0.18	0.184	1.50	2.43	0	719.7	294.5	39.1	3.41	4.06	4.78	1.67	182.7	3.625	1.09	S	1.42	**0.97**	**0.92**	1.51
	PVA 20-15	M	0.84	0.28	0.18	0.184	1.50	2.43	0.15	719.7	294.5	39.1	3.41	4.06	4.78	1.67	205.8	4.083	1.382	S	1.48	**0.99**	**0.95**	1.43
	PVA 20-30	M	0.84	0.28	0.18	0.184	1.50	2.43	0.3	719.7	294.5	39.1	3.41	4.06	4.78	1.67	208.6	4.139	12.62	S	1.39	**0.92**	**0.89**	1.26
Fukumaya and Suwada [13]	MO-1	A	0.4	0.2	0.15	0.16	1.00	0.85	0.845	342	425	78	N/A	N/A	N/A	~0	147	4.9	2.01	F-S				
	PE-1	A	0.4	0.2	0.15	0.16	1.00	0.85	0.845	342	425	55.1	1.9	2.3	N/A	0.6	177	5.9	6.75	F				
	PS-1	A	0.4	0.2	0.15	0.16	1.00	0.85	0.845	342	425	54.5	2.7	3	N/A	1.15	186	6.2	9.35	F				
	MO-2	A	0.4	0.2	0.15	N/A	1.00	0.85	0	342	425	78	N/A	N/A	N/A	~0	92	3.067	1.49	S				
	PE-2	A	0.4	0.2	0.15	N/A	1.00	0.85	0	342	425	55.1	1.9	2.3	N/A	0.6	166	5.533	2.58	F-S				
	PS-2	A	0.4	0.2	0.15	N/A	1.00	0.85	0	342	425	54.5	2.7	3	N/A	1.15	193	6.433	5.6	F-S				
	MO-3	A	0.4	0.2	0.15	N/A	1.00	0.85	0.845	342	425	78	N/A	N/A	N/A	~0	162	5.4	4.02	F				
	PE-3	A	0.4	0.2	0.15	N/A	1.00	0.85	0.845	342	425	55.1	1.9	2.3	N/A	0.6	213	7.1	7.54	F				

[a]tensile strength obtained by 4-point bending test (JCI-S-003-2007).

[b]assumed strength.

In experimental program [10], varied parameters included material (normal concrete [NC] and ECC with 1%, 1.5%, and 2% volume fraction of PVA fibers) and transversal reinforcement ratio (0%, 0.15% and 0.3%). The composite with 1% fiber content essentially did not show pseudo strain-hardening behavior. All specimens in this study failed in shear and thus the failure load corresponds to the shear load capacity of the member. This capacity increased both with increasing reinforcement ratio p_w and tensile strength of cementitious material σ_{max}. In the continuing experimental program [23], the same specimen types were tested. The varied parameters included material (mortar and ECC with 1%, 1.5%, and 2% volume fraction of PVA fibers), transversal reinforcement ratio (0.6%, 0.89%) and strength of longitudinal reinforcement. The main result supports the conclusions of the former program.

R/C Beams with SHCC Layer

Reference [24] reports on experimental investigation of shear behavior of SHCC repaired RC beams. RC beams without stirrups are repaired using SHCC (2% volume fraction of PVA fibers) for the cover thickness (2.5 cm) and twice the cover thickness (5 cm) at the bottom tension section. The former specimen showed higher shear strength and a more stable shear failure behavior than the unrepaired specimens. The later specimen showed similar shear behavior as the plain concrete beam due to the bond failure between the repaired SHCC and original concrete interface.

Short Beam-Columns, Coupling Beams, Flexural Members

Fukyuama and Suwada [13] experimentally tested R/SHCC dampers intended for use for seismic structural control in high-rise buildings. The dampers were configured as short columns exposed to alternating lateral force. The axial elongation as well as rotation of the column end sections were constrained, which induced considerable axial compression to the member at the maximum lateral deflection. Variable parameters in this experiment were the type of cementitious material (MO: mortar, PE: polyethylene fiber SHCC, and PS: steel & polyethylene fiber SHCC) and reinforcement arrangement (1: parallel longitudinal & transversal hoop, 2: X-shaped longitudinal, and 3: X-shaped longitudinal & transversal hoop) – for details see Table 2.3. All type 1 specimens showed cracks inclined at 45° all over the volume. The MO-1 sample eventually failed in compression – by vertical cracks splitting the compressed zone. In PE-1 and PS-1 this failure mode was prevented due to lateral confinement provided by fibers. In type 2 specimens, a diagonal shear crack along the X-shaped reinforcement developed abruptly, causing a drop in load capacity. However, as seen in Table 2.3, in the SHCC dampers this happened at far higher load and deformation angle than in the mortar element. Cracking of type 3 specimens was intermediate between type 1 and 2. All SHCC samples exhibited multiple cracking and superior integrity with steel reinforcement.

Canbolat et al. [25] conducted an experimental study on seismic behavior of coupling beams. The study included reinforced members made of SHCC with polyethylene (PE) or steel fibers with various configurations of conventional reinforcement without and with diagonal bars. The results demonstrated that by using SHCC adequate seismic performance can be achieved even without transversal hoops that are usually required to provide confinement to diagonal rebars. Compared to a typical R/C member with tied diagonal reinforcement (also tested) the SHCC specimen lacking diagonal bars showed lower drift capacity but sustained higher maximum load. The SHCC specimens with diagonal reinforcement showed higher load capacity and sustained drift of at least 4% while retaining about 80% of their shear load carrying capacity.

Parra-Montesinos and Chompreda [26] tested reinforced SHCC cantilever flexural members exposed to displacement reversals. The variable test parameters included type of fiber (PE and steel), fiber volume fraction, type of cementitious matrix, longitudinal reinforcement ratio, and presence or absence of transversal reinforcement. Based on their results, the authors proposed a phenomenological formula for the member shear load capacity, which accounts for the contribution due to SHCC strain-hardening behavior.

Reinforced SHCC Walls and Wall-Frame Assemblies

Assemblies consisting of reinforced PVA-ECC wall surrounded by two R/C columns, R/C top slab and R/C base mat were tested [8]. Alternating horizontal force and constant vertical compression were applied to the top slab, while the base mat was fixed. The test parameters were the wall reinforcement ratio (equal in vertical and horizontal directions), the wall thickness, and anchorage of the wall to the columns. In all of the specimens, the columns exhibited both horizontal bending and inclined shear cracks. The walls were damaged by inclined cracks densely distributed throughout their entire area, with a few interconnecting horizontal cracks. Structural failure occurred in all of the specimens when a shear crack suddenly widened and propagated at the center of the wall.

Fukuyama et al. [27] carried out tests of "ductile wall" using hybrid type SHCC (1% polyethylene fibers and 1% steel code). The ductile wall system shows a unique shape of wall with five short span SHCC columns inside the R/C frame. The ductile walls have additional effective ductility as well as high stiffness and strength of R/C walls.

Low-rise walls with PE and steel fiber SHCC were tested under displacement reversals by Kim and Parra-Montesinos [28]. The authors reported a stable hysteretic behavior up to maximum shear deformation of 2% and excellent damage tolerance.

Beam Column Connections

The experimental study [29] on seismic response of exterior R/C column to steel beam connection involved various designs of the joint, including one constructed with ECC. The ECC connection exhibited a stable load–displacement hysteresis even though it lacked any joint transversal reinforcement. Extensive but distributed diagonal multiple cracking was observed in the joint, but no failure occurred up to the maximum applied story drift of 5%.

The feasibility of using SHCC as means to eliminate the need for transverse confining reinforcement in R/C beam-column connections exposed to seismic loading was experimentally investigated by Parra-Montesinos et al. [30]. The authors reported excellent load and deformation capacity as well as damage tolerance of connections constructed with SHCC with PE fibers. Diagonal multiple cracking was observed within the joint regions, but failure occurred in the adjacent beam plastic hinges. The authors also observed no deterioration in bond of beam longitudinal reinforcement bars passing through the connection.

2.2.2.2 Finite Element Analysis of Shear Failure

Finite element analysis can be employed along with the experimental investigations to clarify the mechanism of shear failure of R/SHCC members. In particular, the issue of effective tensile and compressive strengths of the truss and arch model discussed in Sect. 2.2.3 should be addressed. Since these phenomena are clearly associated with localization of fracture, the FE model must be sufficiently detailed to capture possible initiators of localization: uneven stress distribution due to details at load application points and supports, conventional reinforcement, loading control etc. In addition, the employed constitutive model must reliably represent the material nonlinear behavior, especially that associated with formation, response, and localization of cracks. These requirements are fulfilled by the model proposed by Kabele [31] – see also Sect. 2.3. The model is based on the representation of individual cracks, both when the material undergoes multiple cracking and when fracture localizes. The behavior of each individual crack is characterized by a traction-separation law, that is, by a relation between the bridging traction and relative displacement of the crack surfaces. This relation is established considering that opening and sliding of cracks are resisted by fiber bridging.

The author of ref. [31] numerically simulated experiments on shear beams PVA20-00, PVA20-30 and PVA20-89 described in [23], see also Sect. 2.2.2.1. These beams were produced using ECC with 2% by volume of short PVA fibers (PVA-ECC). The second number indicates the stirrup reinforcement ratio p_w (in hundredths of %). The loading conditions and reinforcement were chosen in all cases such that the members failed in shear. The parameters of the cohesive law for ECC (both for crack opening and sliding) were obtained by fitting the analytical results of beam PVA20-00 to the experimental ones. It is noted, that the composite

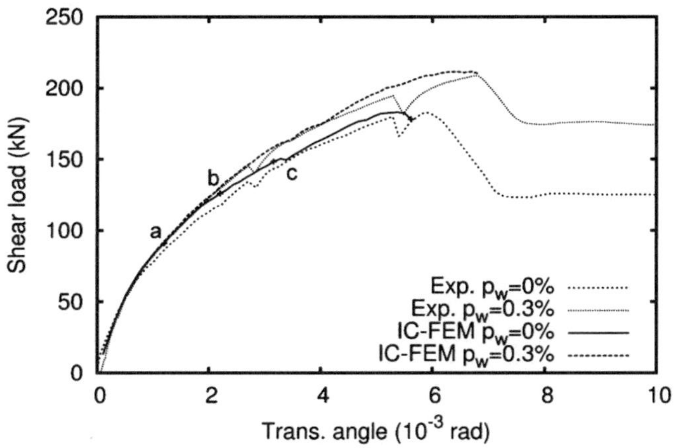

Fig. 2.24 Experimental (Exp.) and analytical (IC-FEM) results of tests on beams with stirrup reinforcement ratios $p_w = 0$ (PVA20-00) and $p_w = 0.3\%$ (PVA20-30). [31]

overall tensile strength and strain capacity corresponding to these parameters (3.02 MPa and 0.7%, resp.) were lower than those determined from direct tension tests on coupon specimens of the same material (4 MPa and 1.6%, respectively). Also, the apparent fiber shear modulus, which was the major parameter controlling sliding resistance of cracks in the used model, had to be set relatively low at 110 MPa (compare with the longitudinal elastic modulus, which is typically 20–40 GPa).

With the above set of parameters, the behavior of beams PVA20-00 and PVA20-30 was very well reproduced, including the prepeak hardening, the peak load and displacement, crack pattern, and the failure mode (Fig. 2.24). Thus, the analysis results can be used to gain a deeper insight into the mechanism of the beams' shear failure. The sequence of cracking computed for beam PVA20-00 is shown in Fig. 2.25. It is seen that fracture initially developed in isolated bands inclined at about 45° and located in the side thirds of the shear span, that is, in the part of the beam that was subjected to the combination of shear force and bending moment. Subsequently, multiple cracking spread into the central part of the beam. Consistent with the experiment, the cracks in the central part were less inclined than those on the sides. The final failure occurred due to localization of fracture into inclined cracks in both side portions of the shear span and evolution of a splitting crack along the reinforcement layer.

Figure 2.26 shows the traction vs. crack opening displacement (δ_n) or crack sliding displacement (δ_t) and stress vs. strain diagrams recorded in three points of the beam marked by the white dots in Fig. 2.25c. The stresses and strains were transformed into local coordinate systems *t*-*n* aligned with cracks in the respective points. From Fig. 2.26c it is obvious that the compressive stress parallel with the cracks remained of a relatively low level, much below the uniaxial compressive strength (which was tested to be 39 MPa). Figure 2.26a indicates that the cracks in

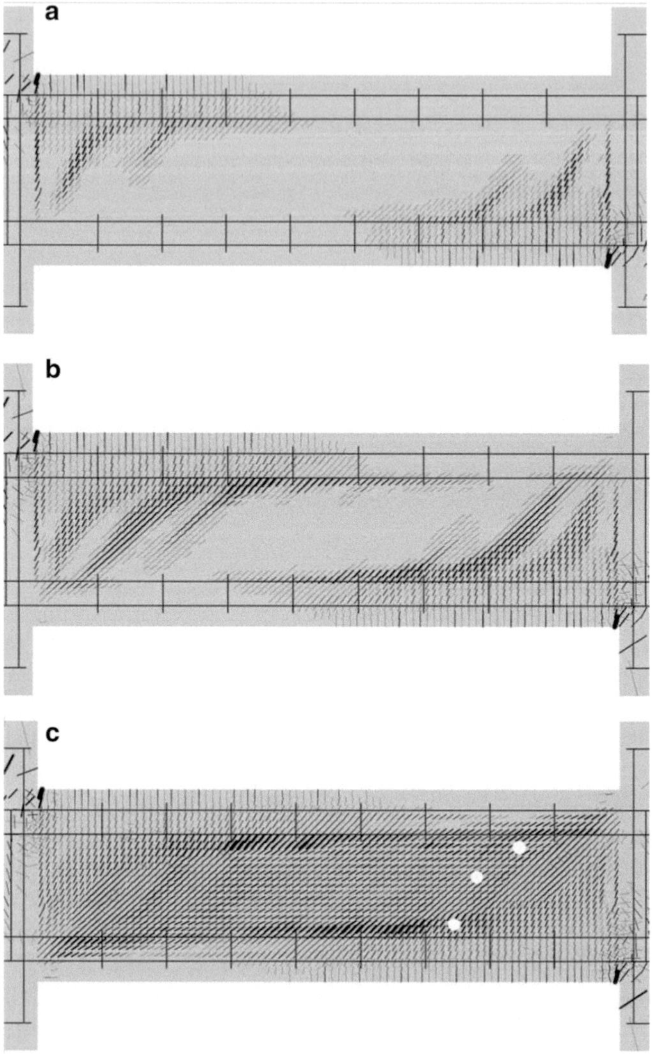

Fig. 2.25 Numerical reproduction of cracking in beam PVA20-00. Phases (**a**) through (**c**) correspond to points marked in Fig. 2.24. [31]

the monitored points did not enter the softening regime even as the structure attained its load capacity. Figure 2.26b shows that cracks exhibited significant sliding and that shear stresses transmitted across cracks may reach similar magnitudes as the normal stresses. In the real experiment, this sliding may have caused damage to fibers bridging cracks since the material lacked coarse aggregate that would otherwise provide matrix interlock. As the model did not explicitly account for fiber

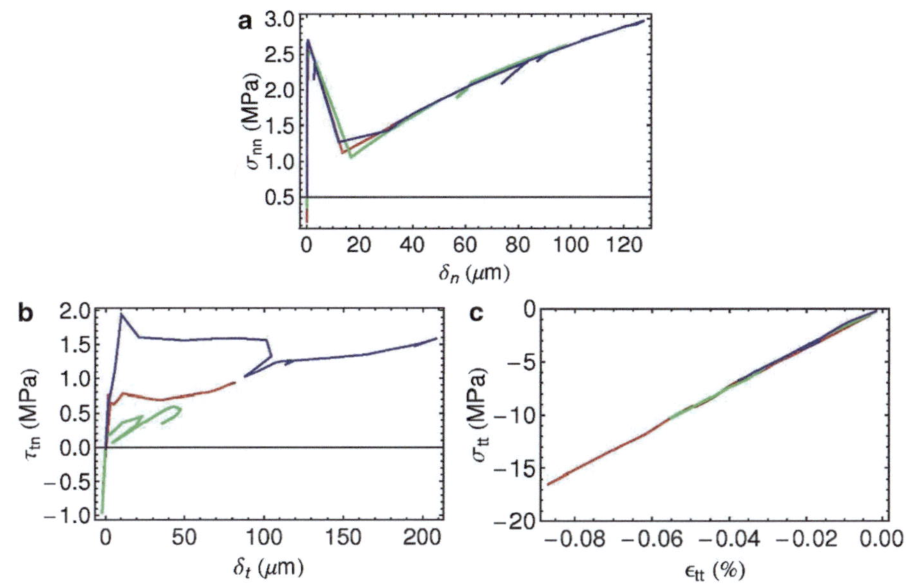

Fig. 2.26 Records of (**a**) normal traction vs. COD, (**b**) tangential traction vs. CSD, (**c**) transversal compressive stress vs. strain calculated at cracks marked by white dots in Fig. 2.25. [31]

damage, this may be the reason why the effective parameters of fiber bridging had to be set lower than what would correspond to the tensile test and why it was necessary to use the low value of the apparent fiber shear modulus.

Figure 2.27 shows the distributions of normal and shear stresses along the beam cross sections at the peak load, together with the corresponding stress distributions derived from the linear elastic solution (Fig. 2.23). It is seen that due to the nonlinear material behavior, the stress distribution considerably deviates from the elastic solution. Therefore, the elasticity solution should not be used to interpret the shear strength of SHCC material from the ultimate load of the Ohno shear beams.

Based on the results of numerical analysis it can be concluded that in R/SHCC structural members, such as shear beams, stress redistribution occurs after cracking. Stress redistribution causes cracks to be exposed not only to opening, but also to shearing. The analyses indicate that under these conditions, SHCC with soft fibers exhibits reduced strength and deformational capacity, compared with its performance in a uniaxial tensile test. Since the material lacks aggregate that would otherwise provide matrix interlock, shearing of cracks may cause excessive loading of bridging fibers, resulting in their rupture and eventually reduced capacity of the material. It should be noted that similar phenomena have been also observed in a recent experimental and analytical work by Suryanto et al. [18, 19].

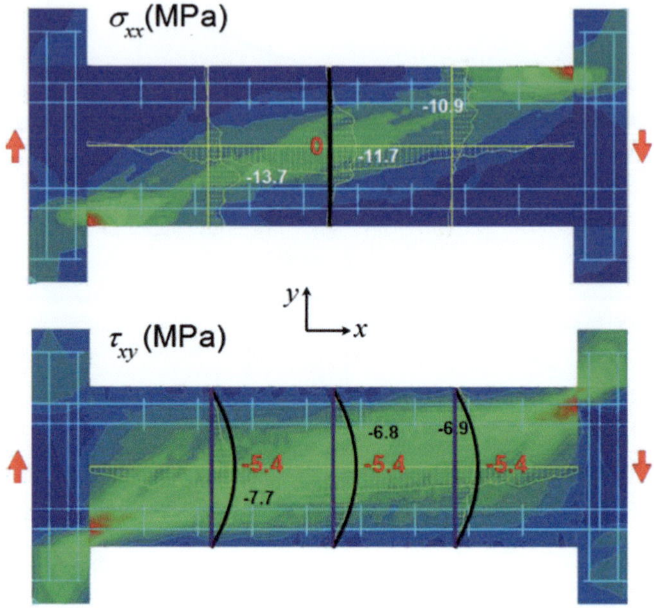

Fig. 2.27 Distributions of normal and shear stresses along the beam cross sections at the peak load (contour bands and *yellow lines*: FEM, *black lines*: elastic solution)

2.2.3 *Evaluation of Shear Capacity of Structural Members*

2.2.3.1 Concept in Evaluation

The shear capacity of R/SHCC structural members such as beams and columns is enhanced when compared with the capacity of ordinary R/C members. It is considered that certain stress is carried through fibers as the bridging effect and / or ductility of tensile and compressive characteristics is improved in the member.

Some proposals to evaluate the shear capacity of R/SHCC members have been reported. In most of them, tensile (crack) strength and / or tensile deformation capacity of SHCC itself are considered, and their effects are taken into the evaluating formulas of shear capacity. For example;

(a) The SHCC effect is substituted for steel reinforcement in R/C evaluation formulas.
(b) Tensile / crack strength of SHCC is added into the truss mechanism.
(c) Shear strength of SHCC is added to R/C evaluation formulas.
(d) Tensile / crack strength of SHCC is added to R/C evaluation formulas.

In the cases of (b) and (d), the direction of tensile stress of SHCC is considered as follows;

1. Same direction with transverse steel reinforcement
2. Perpendicular direction to cracks

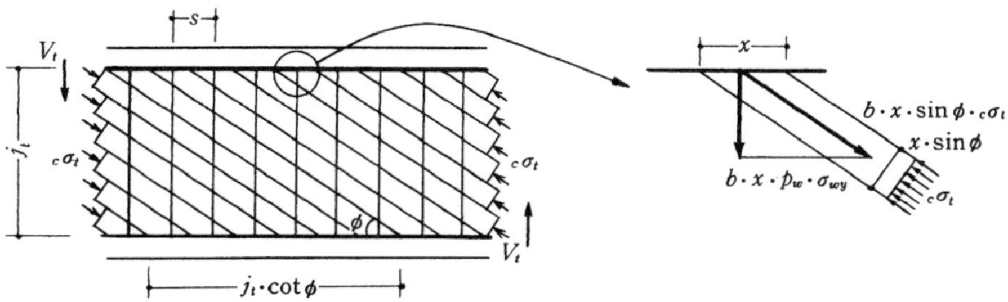

Fig. 2.28 Truss mechanism in AIJ design guideline

3. Direction of principal tensile stress (perpendicular direction to principal compression stress)

It is considered that shear stress exists in the cases of (1) and (2), when the direction of principal tensile stress differs from transverse direction and perpendicular direction of cracks, respectively.

Some proposed formulas stands on the basis of the arch – truss mechanism which is originally adopted in "Design Guideline for Earthquake Resistant Reinforced Concrete Buildings Based on Ultimate Strength Concept (Architectural Institute of Japan)" [32]. In this section, the original formulas to evaluate the shear capacity of R/C members are introduced.

The original formulas had been developed on the basis of plasticity theory which assumes that the shear capacity can be expressed by superposition of truss mechanism and arch mechanism. Figure 2.28 shows the concept of truss mechanism. The ϕ is the angle of compressive strut of concrete which has equilibrium with tensile force of transverse reinforcements. The value of $\cot\phi$ should be the range from 1.0 to 2.0 to have the maximum value of shear force provided by truss mechanism comparing with that of arch mechanism. Shear capacity by truss mechanism (V_t) is given by Eq. (2.3). Compressive stress in the truss strut ${}_c\sigma_t$ is given by Eq. (2.4).

$$V_t = b \cdot j_t \cdot p_w \cdot \sigma_{wy} \cdot \cot\phi \tag{2.3}$$

$${}_c\sigma_t = (1 + \cot^2\phi) \cdot p_w \cdot \sigma_{wy} \tag{2.4}$$

where,

b : width of member
j_t : distance between tensile and compressive longitudinal reinforcement
p_w : transverse reinforcement ratio
σ_{wy} : yield strength of transverse reinforcement
ϕ : angle of compressive strut (truss)

Figure 2.29 shows the concept of arch mechanism. If we have the assumption that the angle θ, which expresses the angle of compressive strut of arch mechanism, changes to have the maximum value of shear force provided by arch mechanism,

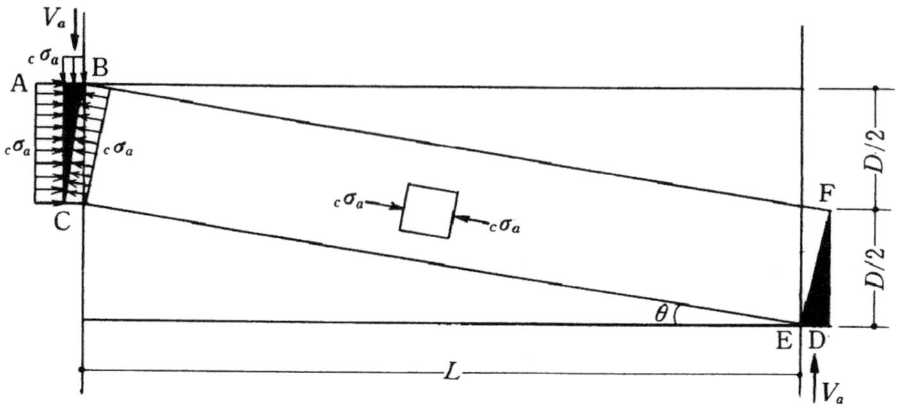

Fig. 2.29 Arch mechanism in AIJ design guideline

the θ is given by Eq. (2.5). The shear capacity by arch mechanism (V_a) is given by Eq. (2.6). The compressive stress of arch strut is expressed as $_c\sigma_a$.

$$\tan\theta = \sqrt{(L/D)^2 + 1} - (L/D) \tag{2.5}$$

$$V_a = {}_c\sigma_a \cdot \tan\theta \cdot b \cdot (D/2) \tag{2.6}$$

where,

L : length of member (anti-symmetrical bending moment)
D : depth of member
θ : angle of compressive strut (arch)

Neglecting the difference of compressive strut angle between truss and arch mechanism, compressive failure at strut concrete occurs as given by Eq. (2.7). Where, ν indicates the reduction factor for compressive strength of concrete which has cracks and deformation. The value of ν is originally given by Eq. (2.8). However, Eq. (2.9) is often used for high strength concrete.

$$\nu \cdot \sigma_B = {}_c\sigma_t + {}_c\sigma_a \tag{2.7}$$

$$\nu = 0.7 - \sigma_B/2000 \ \left(\text{in kgf/cm}^2\right) \tag{2.8}$$

$$\nu = 1.7\sigma_B{}^{-0.333} \ (\text{in MPa}) \tag{2.9}$$

where,

ν : reduction factor for concrete compressive strength
σ_B : concrete compressive strength

Introducing the β, ratio of compressive stress of truss strut to reduced concrete strength as Eq. (2.10), leads total shear capacity by the superposition of truss

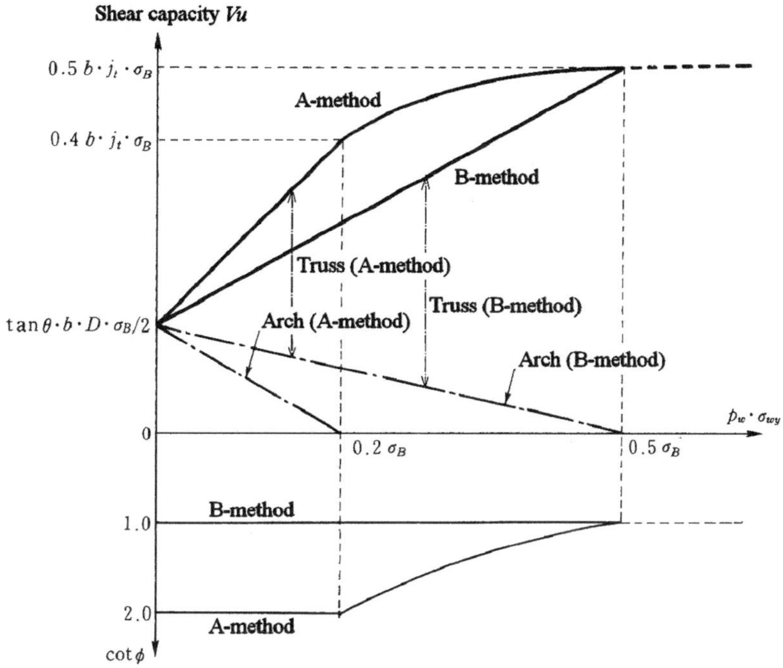

Fig. 2.30 Relative components of truss and arch mechanism

mechanism and arch mechanism as given by Eq. (2.11). The value of cotϕ is to be the minimum value among Eqs. (2.12)–(2.14).

$$\beta = \frac{(1 + \cot^2\phi) \cdot p_w \cdot \sigma_{wy}}{\nu \cdot \sigma_B} \tag{2.10}$$

$$V = b \cdot j_t \cdot p_w \cdot \sigma_{wy} \cdot \cot\phi + \tan\theta \cdot (1 - \beta) \cdot \nu \cdot \sigma_B \cdot b \cdot D/2 \tag{2.11}$$

$$\cot\phi = 2.0 \tag{2.12}$$

$$\cot\phi = j_t/(D \cdot \tan\theta) \tag{2.13}$$

$$\cot\phi = \sqrt{\frac{\nu \cdot \sigma_B}{p_w \cdot \sigma_{wy}} - 1} \tag{2.14}$$

Due to these procedures, shear capacity increases as the capacity provided by truss mechanism also increases with increment of the value of $p_w\sigma_{wy}$. On the other hand, the shear capacity provided by arch mechanism decreases as shown in Fig. 2.30. These formulas are called as "A-method". In "B-method", cotϕ is set to the constant value of 1.0 (ϕ is equal to 45°).

2.2.3.2 Examples of Design Formulas

Truss Mechanism Based Formula

Nagai proposed a shear design formula, which is categorized into type (b-3) [33]. The tensile strength of SHCC is taken into account as part of the truss mechanism. However, the tensile strength is to be half of the yield strength (almost the same value as the first cracking strength) for structural design calculation, and the value of cotϕ is fixed to 1.0.

$$V = b \cdot j_t \cdot (p_w \cdot \sigma_{wy} + \sigma_t) \cdot \cot\phi + \tan\theta \cdot (1 - \beta) \cdot \nu \cdot \sigma_B \cdot b \cdot D/2 \tag{2.15}$$

$$\beta = \frac{(1 + \cot^2\phi) \cdot p_w \cdot \sigma_{wy} + \cot^2\phi \cdot \sigma_t}{\nu \cdot \sigma_B} \tag{2.16}$$

$$\cot\phi = 1.0 \tag{2.17}$$

$$\nu = 1.7{\sigma_B}^{-0.333} (\text{in MPa}) \tag{2.18}$$

where,

σ_t: $\sigma_{ty}/2$

σ_{ty}: yield strength of SHCC (lowest tensile stress after first cracking).

Tensile Contribution Super-Imposing Formula

Shimzu and Kanakubo also proposed a formula categorized into type (d-3) [34]. This formula takes into account the tensile strength of SHCC as part of the total shear resistance mechanism. The tensile strength of SHCC is reduced by a reduction factor, which is assumed from experiment results.

$$V = b \cdot j_t \cdot p_w \cdot \sigma_{wy} \cdot \cot\phi + \tan\theta \cdot (1 - \beta) \cdot \nu \cdot \sigma_B \cdot b \cdot D/2 + b \cdot j_t \cdot \nu_t \cdot \sigma_t \cdot \cot\phi \tag{2.19}$$

$$\beta = \frac{\left(1 + \cot^2\phi\right) \cdot p_w \cdot \sigma_{wy}}{\nu \cdot \sigma_B} \tag{2.20}$$

$$\cot\phi = \min\left\{2, \sqrt{\frac{\nu \cdot \sigma_B}{p_w \cdot \sigma_{wy}} - 1}\right\} \tag{2.21}$$

$$\nu = 1.7{\sigma_B}^{-0.333} (\text{in MPa}) \tag{2.22}$$

where,

σ_t : tensile strength of SHCC

ν_t : reduction factor for tensile strength of SHCC (=0.41)

Tensile Contribution Super-imposing Formula – JSCE method

JSCE's proposal [35] is categorized into type (d-2). The tensile strength of SHCC is taken into account as part of the whole shear resistance mechanism. The tensile yield strength of SHCC (almost the same value as the first cracking strength) is utilized.

$$V = V_c + V_s + V_f \tag{2.23}$$

$$V_c = \beta_d \cdot \beta_p \cdot \beta_n \cdot f_{vc} \cdot b \cdot d \tag{2.24}$$

$$V_s = (A_w \cdot \sigma_{wy}/s_s) \cdot z \tag{2.25}$$

$$V_f = (f_v/\tan \beta_u) \cdot b \cdot z \tag{2.26}$$

$$f_{vc} = 0.7 \times 0.20\ \sqrt[3]{\sigma_B}\ \text{(in MPa)} \tag{2.27}$$

$$\beta_d = \sqrt[4]{1/d}\ \text{(in m)} \tag{2.28}$$

$$\beta_p = \sqrt[3]{100 \cdot p_w} \tag{2.29}$$

$$\beta_n = 1 + M_0/M_d\ (\leq 2) \tag{2.30}$$

where,

V_c : shear capacity without transverse reinforcement
V_s : shear capacity of transverse reinforcement
V_f : shear capacity of fiber
d : effective depth
A_w : cross sectional area of a pair of transverse reinforcement
s_s : spacing of transverse reinforcement
z : distance from location of compressive stress resultant to centroid of tensile reinforcement (=d/1.15)
f_v : tensile yield strength of SHCC (lowest tensile stress after first cracking)
β_u : angle of diagonal crack surface to the member axis (=45°)
M_0 : bending moment necessary to cancel stress due to axial force at extreme tension fiber corresponding to bending moment M_d
M_d : design bending moment

Modified Truss Mechanism Based Formulas

In reference [36], the following method was proposed. In Fig. 2.31, the truss mechanism and arch mechanism are defined as the shear resistance mechanism

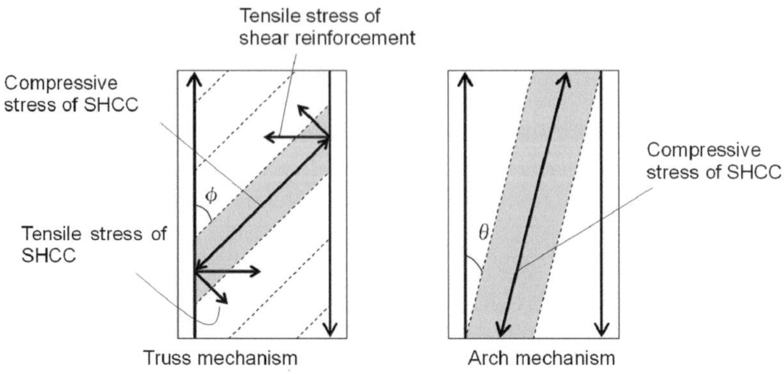

Fig. 2.31 Definition of truss mechanism and arch mechanism

of SHCC members. Furthermore, in Fig. 2.32, the following five failure modes are defined to predict the shear strength considering the state of each constitutive material.

Mode 1 : Shear reinforcements do not yield in the truss mechanism but before the tensile principal strain of SHCC reaches ultimate tensile strain capacity a compressive strut reaches a compressive strength that results in destruction. Shear resistance by the arch mechanism is not produced.

Mode 2 : Before a compressive strut reaches compressive strength in the arch mechanism, a shear reinforcement yields in the truss mechanism, and the tensile principal strain of SHCC reaches ultimate tensile strain capacity and results in destruction.

Mode 3 : Shear reinforcements yield in a truss mechanism and before the tensile principal strain of SHCC reaches ultimate tensile strain capacity, in the arch mechanism, a compressive strut reaches a compressive strength that results in destruction.

Mode 4 : Shear reinforcements yield in the truss mechanism and before the tensile principal strain of SHCC reaches ultimate tensile strain capacity, a compressive strut reaches a compressive strength that results in destruction. Shear resistance by the arch mechanism is not produced.

Mode 5 : Shear reinforcements yield in the truss mechanism and after the tensile principal strain of SHCC reaches ultimate tensile strain capacity, a compressive strut reaches a compressive strength that results in destruction. Shear resistance by the arch mechanism is not produced.

The Fafitis-Shah model is used for the compressive stress–strain relationship of SHCC, because it is confirmed that the value of the formula agrees well with the experimental value.

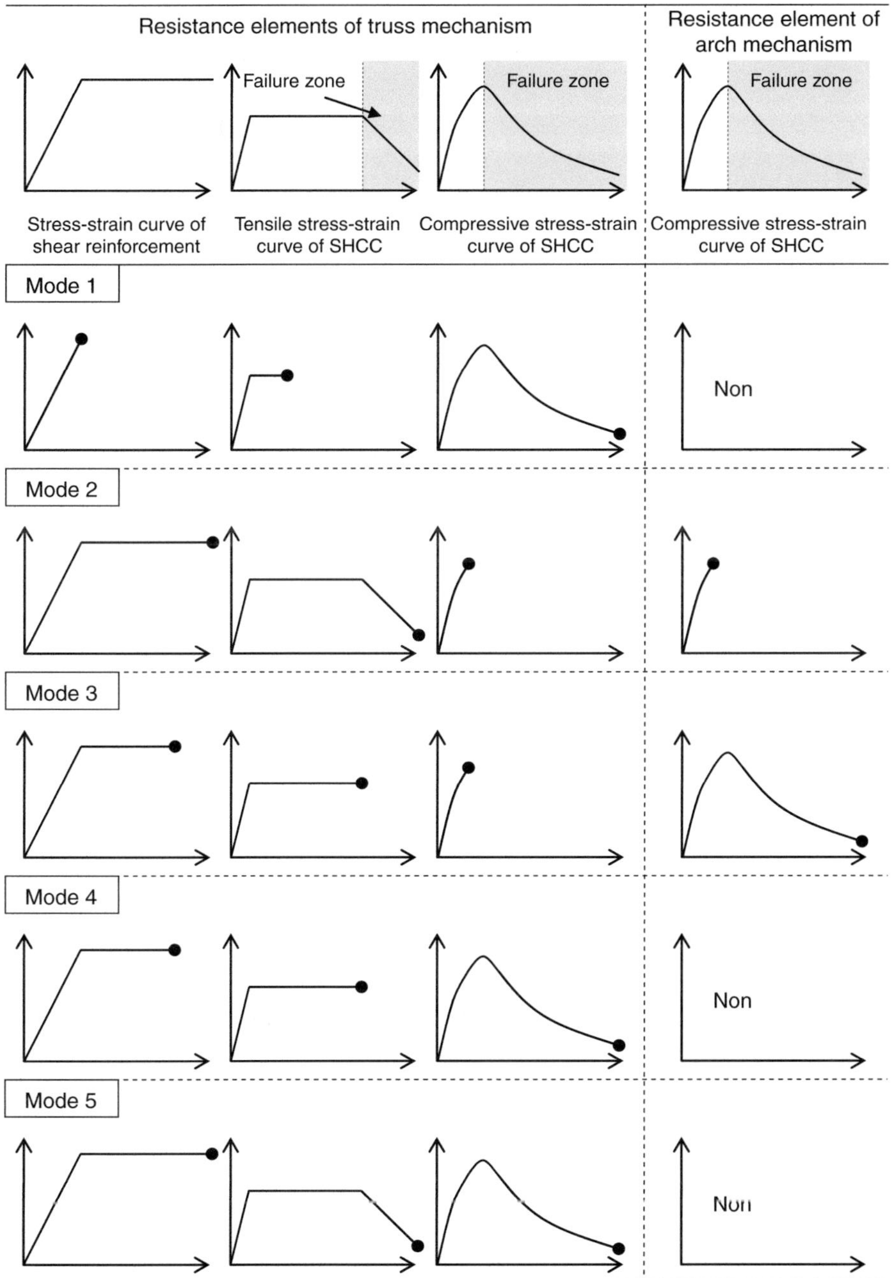

Fig. 2.32 Definition of failure mode

$$\sigma_c = \nu\sigma_B \left\{ 1 - \left(1 - \frac{\varepsilon}{\varepsilon_0} \right)^A \right\} \tag{2.31}$$

$$A = E_c \frac{\varepsilon_0}{\nu\sigma_B} \tag{2.32}$$

where, σ_B : compressive strength, ν : effective compressive-strength multiplier, ε_0 : strain at the time of a compressive strength, and E_c : Young's modulus.

The tensile stress–strain relationship of SHCC is presented by the bi-linear model that disregards the stress rise during multiple cracking sequence, and where E_c is the Young's modulus, σ_t is the crack generating stress (equivalent to yield stress), and ε_{PSH} is the ultimate tensile strain. In addition, E_c, ε_0, σ_t, and ε_{PSH} of SHCC, should be preferably obtained by an appropriate calculation method. However, since there was no calculation formula that suitably took into consideration the influence of the mixing factor of SHCC, the material-test value was used directly.

The stress–strain relationship of shear reinforcement is taken as a bi-linear model that disregards the stress rise after yielding and where E_{ws} is the Young's modulus and σ_{wy} is the yield stress.

The load and deformation of SHCC members at the time of shear failure are calculated by the following equations.

$$Q_u = b j_t \tau_{truss} + bD\tau_{arch} \tag{2.33}$$

$$\delta_u = L\gamma \tag{2.34}$$

where, b : member width, j_t : effective depth of member, D : whole depth of member, τ_{truss} : carrying shear stress of truss mechanism, τ_{arch} : carrying shear stress of arch mechanism, L : member length and γ : shear strain.

In addition, τ_{truss}, τ_{arch}, and γ are obtained in accordance with the value of $\cot\phi$.

As a deformation component of each mechanism to obtain shear strain, deformation in the truss mechanism is defined as compressive deformation component (component A) and tensile deformation component (component B) as shown in Fig. 2.33. Deformation in the arch mechanism is defined as shown in Fig. 2.34.

$$\cot\phi = \min\left(2.0, \frac{j_t}{D\tan\theta}, \sqrt{\frac{\nu\sigma_B - p_w\sigma_{wy}}{p_w\sigma_{wy} + \sigma_t}} \right) \geq 1.0 \tag{2.35}$$

$$\tan\theta = \sqrt{\left(\frac{L}{D}\right)^2 + 1} - \left(\frac{L}{D}\right)$$

$$\nu = 0.61 \tag{2.36}$$

where, ϕ: compressive strut angle of truss mechanism, θ: compressive strut angle of arch mechanism, p_w : shear reinforcement ratio, σ_{wy} : yield strength of shear reinforcement, σ_t : crack strength of SHCC.

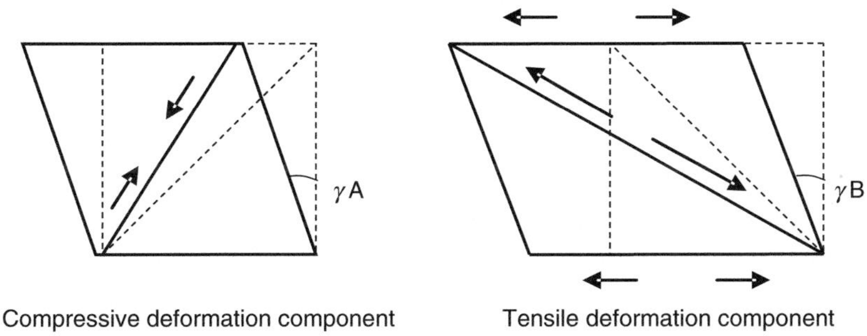

Fig. 2.33 Definition of deformation component in truss mechanism

Fig. 2.34 Definition of deformation component in arch mechanism

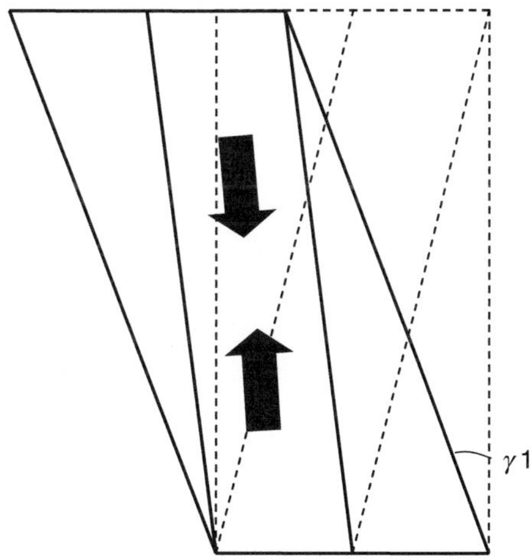

(a) **cotϕ = 1.0**

When the calculation value by Eq. (2.35) becomes cot$\phi = 1.0$, the amount of the shearing reinforcement is large. The failure mode becomes mode 1 that the compressive strut of the truss mechanism reaches $\nu\sigma_B$, and the maximum strength is calculated by the following equations.

$$\sigma_{1,truss} = \nu\sigma_B \quad (2.37)$$

$$\gamma_A = \frac{2\varepsilon_0}{\sin 2\phi} = 2\varepsilon_0 \quad (2.38)$$

$$\gamma_B = \frac{\nu\sigma_B \sin^2\phi - \sigma_t \cos^2\phi}{p_w E_{ws} \cot\phi} \quad (2.39)$$

$$\gamma_{truss} = \gamma_A + \gamma_B \quad (2.40)$$

$$\tau_{truss} = \frac{\nu\sigma_B}{2} \quad (2.41)$$

$$\tau_{arch} = 0 \tag{2.42}$$

(b) **cotϕ = 2.0**

When the calculation value by Eq. (2.35) becomes cotϕ = 2.0, the amount of shearing reinforcement is little. The failure mode becomes mode 2, where the tensile strut of the truss mechanism reaches ε_{PSH}, or, it becomes mode 3, where the compressive strut of the arch mechanism reaches $\nu\sigma_B$. Mode 2 and mode 3 have the stress of the arch mechanism. The following two smaller shear strains reach the shear strength.

- Shear strain γ_{arch} when the compressive strut reaches the compressive strength under the arch mechanism.
- Shear strain γ_{truss} when the tensile strut reaches the ultimate tensile strain under the truss mechanism.

Calculation of γ_{arch}

$$\gamma_{arch} = \frac{2\varepsilon_0}{\sin 2\theta} \tag{2.43}$$

Calculation of γ_{truss}

$$\gamma_B = \frac{\varepsilon_{PSH} \cos\phi}{\cot\phi} \tag{2.44}$$

$$\gamma_A = \frac{2\varepsilon_0}{\sin 2\phi}\left\{1 - \left(1 - \frac{p_w\sigma_{wy} + \sigma_t\cos^2\phi}{\nu\sigma_B\sin^2\phi}\right)^{\frac{1}{A}}\right\} \tag{2.45}$$

$$\gamma_{truss} = \gamma_A + \gamma_B \tag{2.46}$$

Judgment of failure mode

$$\gamma = \min\{\gamma_{arch}, \gamma_{truss}\} \tag{2.47}$$

Calculation of shear strength

$$\gamma_{\text{arch}} < \gamma_{\text{truss}} (\text{Mode 3})$$

$$\tau_{truss} = p_w\sigma_{wy}\cot\phi + \sigma_t\cos^2\phi \tag{2.48}$$

$$\tau_{arch} = \frac{\tan\theta\,(1-\beta)\,\nu\sigma_B}{2} \tag{2.49}$$

$$\beta = \frac{p_w\sigma_{wy}\left(1+\cot^2\phi\right) + \sigma_t\cot^2\phi}{\nu\sigma_B} \tag{2.50}$$

$$\gamma_{\text{arch}} > \gamma_{\text{truss}}(\text{Mode 2})$$

$$\tau_{truss} = p_w \sigma_{wy} \cot\phi + \sigma_t \cos^2\phi \tag{2.51}$$

$$\varepsilon_{1,arch} = \gamma_{truss} \sin\theta \cos\theta \tag{2.52}$$

$$\sigma_{1,arch} = \nu\sigma_B \left\{ 1 - \left(1 - \frac{\varepsilon_{1,arch}}{\varepsilon_0} \right)^A \right\} \tag{2.53}$$

$$\tau_{arch} = \frac{\tan\theta\,(1-\beta)\,\sigma_{1,arch}}{2} \tag{2.54}$$

(c) **1.0 < cotϕ < 2.0**

When the calculation value obtained by Eq. (2.35) becomes 1.0 < cotϕ <2.0, the amount of shearing reinforcement is somewhat large. The failure mode is either mode 4, where destruction occurs when the compressive strut of the truss mechanism reaches $\nu\sigma_B$, or mode 5, where destruction occurs when the compressive strut reaches $\nu\sigma_B$ by maintaining the resistance of the shear reinforcement even after the tension strut of the truss mechanism reaches ε_{PSH}. Distinction between these two failure modes depends on ε_{PSH} Hence, cotϕ should not be used as the distinction criterion for these two modes, but ε_{PSH} can be used. Based on the results of FEM analysis of cases in which the failure mode became mode 5, the $\varepsilon_{PSH} = 1.0\%$ is used as the boundary value for failure mode distinction.

When $\varepsilon_{PSH} > 1.0\%$,Mode 4
When $\varepsilon_{PSH} < 1.0\%$,Mode 5

$$\gamma_A = \frac{2\varepsilon_0}{\sin 2\phi} \tag{2.55}$$

$$\gamma_B = \frac{\varepsilon_w}{\sin 2\phi} \tag{2.56}$$

$$\varepsilon_w = \left(0.01\cos\phi - \varepsilon_{wy}\right)(\cot\phi - 1) + \varepsilon_{wy} \tag{2.57}$$

$$\gamma_{truss} = \gamma_A + \gamma_B \tag{2.58}$$

When $\varepsilon_{PSH} > 1.0\%$,

$$\tau_{truss} = p_w \sigma_{wy} \cot\phi + \sigma_t \cos^2\phi \tag{2.59}$$

When $\varepsilon_{PSH} < 1.0\%$,

$$\tau_{truss} = p_w \sigma_{wy} \cot\phi \tag{2.60}$$

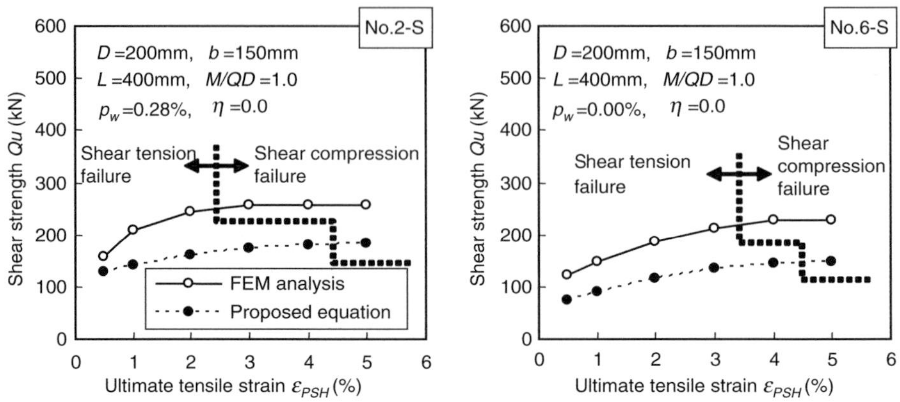

Fig. 2.35 Example comparing calculation values by proposed equation and FEM analysis

$$\tau_{arch} = 0 \tag{2.61}$$

An example comparing the calculation values by the proposed equation and the FEM analysis for the relation of maximum strength Q_u and ε_{PSH} is shown in Fig. 2.35. The change of Q_u according to the change of ε_{PSH} is approximately reproduced according to this figure although the calculation value by the proposed equation reproduces the result of the FEM analysis [56] considerably on the safety side. However, differences are observed in the failure mode. These differences may be due to the fact that although the failure mode is judged based on the compressive strut angle ($\cot\phi = 1.0, 1.0 < \cot\phi < 2.0, \cot\phi = 2.0$) of the truss mechanism, it does not provide an accurate measure of changes in failure mode owing to ε_{PSH}.

2.2.4 Concluding Remarks

The experimental and analytical results presented in this report indicate that there are substantial differences in the shear resisting mechanisms of R/C and R/SHCC members. In order to develop a reliable yet simple model for prediction of the shear load carrying capacity of R/SHCC members, this mechanism should be further investigated under various test configurations (reinforcement ratios, shear-span to depth ratio, etc.). To this end, finite element analysis can be efficiently used. Consequently, the extensive source of existing experimental data cited in this paper can be utilized to validate the model.

(P. Kabele, T. Kanakubo and H. Suwada)

2.3 Finite Element Modeling

2.3.1 Introduction

Finite element (FE) modeling of SHCC materials and structures has mostly focused at the following two objectives: reproduction and prediction of the specific fracture phenomena that occur at the scale of a material sample (micro- to mesoscale) and analysis of structural members/structures made of SHCC (macro- to structural scale). The first category is represented, for example, by work of Dick-Nielsen et al. [37, 38], who used a FE model to examine the effects of specimen geometry and boundary conditions on the process of multiple cracking in a uniaxial tensile test. The second category is of the main interest in the present chapter, since these models can serve as tools for structural design and evaluation of structural performance of SHCC. Structural FE analysis is also an important aid for practical implementation of the Integrated Structures and Materials Design (ISMD) concept [39], which implies a two-way interaction between structural design and material development. Employing ISMD, structures are to be conceptually designed so as to take full advantage of material properties, while materials are to be tailored to specific structural needs.

For a structural FE analysis to fulfill the above objectives, the underlying constitutive model and its implementation into an FEM framework should, in particular, properly capture the specific mechanical phenomena of SHCC materials: distributed multiple cracking as well as subsequent localization of damage and formation and behavior of distinct large cracks. Furthermore the models should be valid for:

1. Arbitrary stress state (tension, compression and their combination),
2. Arbitrary loading path at a material point (esp. non-proportional change of stress components, change of principal stress direction)
3. Unloading, load reversals, cyclic loading, fatigue, creep, etc.

It is noted that most models presently available in the literature do not aspire to fully meet all these criteria; some models are more biased toward accurately representing the behavior under multiaxial stress state (criteria 1 and 2) while others focus on the cyclic behavior or creep derived mostly from a uniaxial response (criterion 3). In the forthcoming text we review some of the existing models.

2.3.2 Continuum Based Models

The models reviewed in this section adopt a phenomenological approach to represent the SHCC material in multiple-cracking state. A cracked composite is viewed as a continuum whose nonlinear deformation due to cracking is represented by means of inelastic strain.

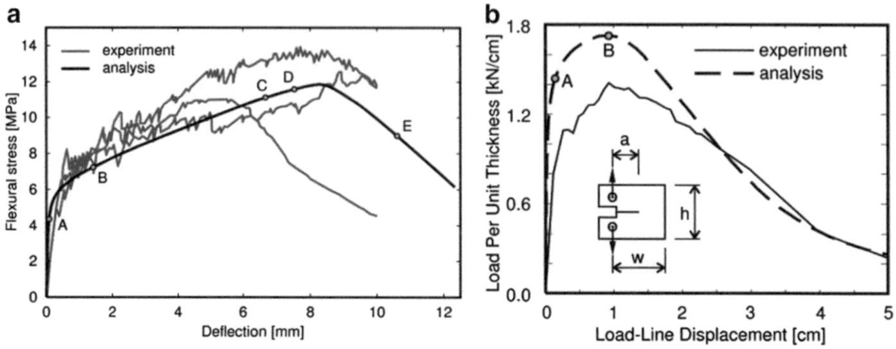

Fig. 2.36 Response of 4-point bending beam (**a**) and compact tension test specimen (**b**) calculated with plasticity-based model [40]

Kabele and Horii [40, 41] developed a plasticity-based model for two-dimensional finite element analysis of structures made of SHCC materials. A similar approach was also used by Kullaa [42]. The model combines the incremental theory of plasticity with multiple yield functions and hardening rules for a material in the multiple cracking state, and a discrete crack modeling concept for localized cracks. The condition for initiation of pseudo strain-hardening behavior associated with tensile multiple cracking is represented by the Rankine yield function. Deformation resulting from multiple cracking is treated as an inelastic cracking (plastic) strain. The anisotropy induced by multiple cracking is reflected by using the kinematic hardening rule. Once the composite tensile strain capacity is attained in any element and tension softening starts, a discrete crack is introduced in the direction perpendicular to the maximum principal cracking strain. The discrete crack is implemented by elements with embedded displacement discontinuity. The material nonlinear behavior in compression was treated by von Mises plasticity with isotropic hardening. The inelastic strain associated with the von Mises yield function is separate from the cracking strain. As the input, the model accepted the hardening portion of the uniaxial tensile stress–strain relationship, the tension softening relationship and the compressive stress–strain curve. The model was validated by successful reproduction of multiple and localized cracking in bending beams (Fig. 2.36a) and by reproduction of cracking observed in compact tension tests. The model, however, exhibits a tendency to overestimate stiffness and strength of the compact tension test specimens (Fig. 2.36b) as well as of structural elements failing in shear, such as shear beams.

Boshoff and van Zijl [44] proposed and implemented in FEM a constitutive model for cracked SHCC based on the isotropic damage formulation. The model treats both the distributed multiple cracking as well as subsequent fracture localization associated with onset of softening behavior. The authors employ a modified definition of equivalent strain, which makes it possible to capture shear-dominated failure. The model of the softening material is regularized by incorporating a localization length scale, which is related to the phenomenon of fiber pullout. Subsequently, the finite element solution is mesh-size independent. The model was verified by reproducing experiments failing in bending and shear.

Van Zijl and Boshoff [45] developed and implemented in FEM program DIANA a constitutive model for rate-dependent behavior of SHCC. The model is based on multisurface continuum plasticity and incorporates viscoelasticity for bulk creep. The plasticity term, which uses the Rankine yield function, represents cracking. The cracking rate term accounts for cracking viscosity and delayed fiber pullout. The model was shown to capture rate-enhanced tensile resistance and creep fracture under uniaxial tension.

2.3.3 Smeared Crack and Homogenization Based Models

This section is devoted to constitutive models for SHCC materials based on the smeared crack concept or homogenization of continuum with cracks.

Takeuchi and Kabele [43, 46] adapted the fixed smeared crack model included in general purpose FE program MARC [47] for SHCC materials. In the crack-normal direction, linear hardening relation between stress and cracking strain is used to describe the material behavior during the multiple-cracking regime. Linear softening, defined in terms of stress vs. crack opening displacement, is used for localized cracks. The authors recognize that crack sliding must be properly treated in order to capture behavior of structural elements under complex loading conditions, which result in non-proportional stress paths at different material points. Hence it is considered that crack sliding is resisted by bridging fibers, which are modeled as short elastic beams. The resulting overall shear stiffness depends on the crack opening, which was implemented by a variable shear retention factor. The model was validated by reproduction of experiments on the Ohno shear beams. Nevertheless, the analytical results still showed somewhat stiffer response than the experiment unless the crack shear (sliding) stiffness was considerably reduced (Fig. 2.37). The model was then used to predict the performance of ECC panels intended for seismic retrofit, ECC overlays undergoing shrinkage or ECC-based repair of spalled concrete [43, 48].

Kabele [49, 50] established a systematic framework for equivalent-continuum based modeling for SHCC in multiple cracking state by utilizing the methods of micromechanics of solids with defects. The constitutive law is obtained as a relationship between overall stress and overall strain of a representative volume element (RVE). The RVE is chosen as a volume which contains a large number of diffuse cracks, yet, compared to the scale of the analyzed structural member can be regarded as a material point. The RVE is modeled as a solid element throughout intersected by up to three mutually orthogonal sets of fiber-bridged cracks. It is assumed that the first set of cracks occurs when the maximum principal stress reaches the first-crack strength; the onset of subsequent orthogonal cracks is controlled by the normal stress in the respective directions. The cohesive

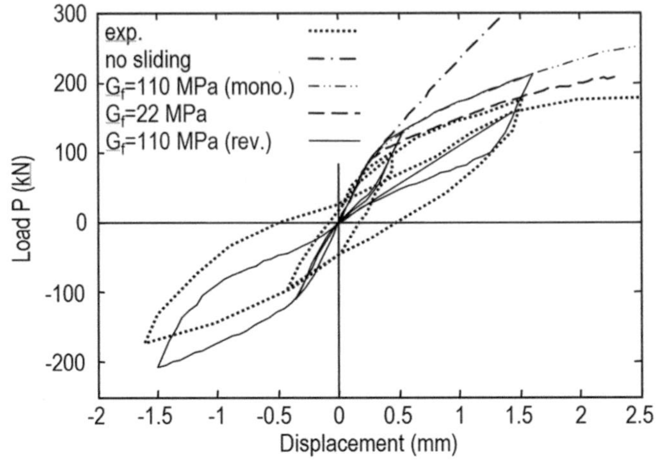

Fig. 2.37 Response of Ohno shear beam calculated by Kabele et al. [43]

relationship on individual cracks is derived separately for crack-normal (opening) and crack-tangent (sliding) direction. The overall stress–strain relation in crack-normal direction is calibrated experimentally. In the crack-tangent direction, it is assumed that the bridging action is provided by the protruding parts of fibers crossing a crack, which are modeled as short elastic beams (in a similar way as in the earlier model [43, 46]). The equivalent-continuum based model was also used to investigate fracture behavior of Ohno shear beams [58].

Han et al. [51] extended the total strain-based rotating crack model originally proposed for concrete by Feenstra (cited in [51], p. 750) for analysis of R/ECC columns under cyclic bending load. The material response in directions perpendicular and parallel with cracks was modeled by a stress function, which involved a total strain and an internal variable that recorded the history of loading and unloading. The stress function was based on experimental data obtained from uniaxial reversed cyclic load tests. The authors studied the effect of accounting for bond slip or not, reducing the initial stiffness of SHCC (assumed due to shrinkage cracking) and varying the unloading/reloading response of SHCC. As seen in Fig. 2.38, despite reducing the initial material stiffness and accounting for bond-slip, the analyses predicted higher initial stiffness of the structure than what was observed in experiments. Also, the computed hysteretic energy dissipation was larger. However, the ultimate load and drift after several cycles matched the experimental results well. The authors found that modeling of bond-slip has more importance in capturing the global experimental response than does the assumed initial stiffness of SHCC.

The constitutive model proposed by Han et al. [51] was also used by Kesner and Billington [52] to analyze R/ECC in-fill panels. As it is obvious in Fig. 2.39, the simulation was able to reproduce reasonably well the load capacity and residual

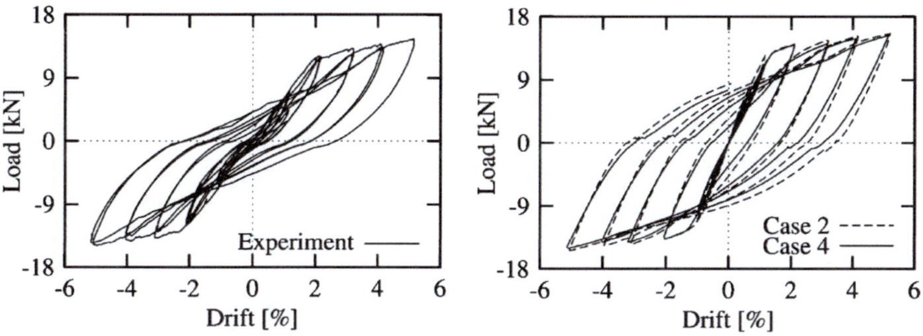

Fig. 2.38 Response of cantilever column under bending load presented by Han et al. [51]: (**a**) experiment (**b**) simulation with perfect bond vs. bond-slip and SHCC modulus reduced to 60%

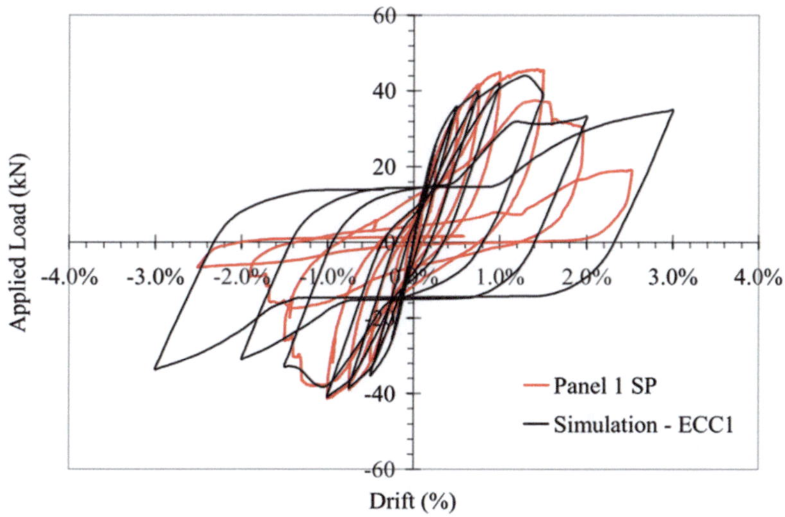

Fig. 2.39 Response of R/ECC in-fill panel calculated by Kesner and Billington [52]

displacement of the panel up to about 2% drift. At higher drift levels, the residual load carrying capacity of the panel was overestimated, which the authors attributed to considering perfect bond and neglecting fracture of reinforcing steel bars.

Dick-Nielsen et al. [37] developed and implemented in FEM a plasticity-based damage-mechanics model for SHCC materials. The model is based on the fixed multiple smeared crack approach. In contrast to other models, cohesive effects on cracks are attributed not only to fibers, but also to matrix interlock. The matrix crack-bridging effect is described using a plasticity damage model. The model is therefore able to capture the dilation effect that occurs during mode II crack opening (sliding). The model provides information about crack opening orientation and spacing. Capabilities of the model have been demonstrated by a simulation of material response to non proportional loading and reproduction of a SHCC panel [37].

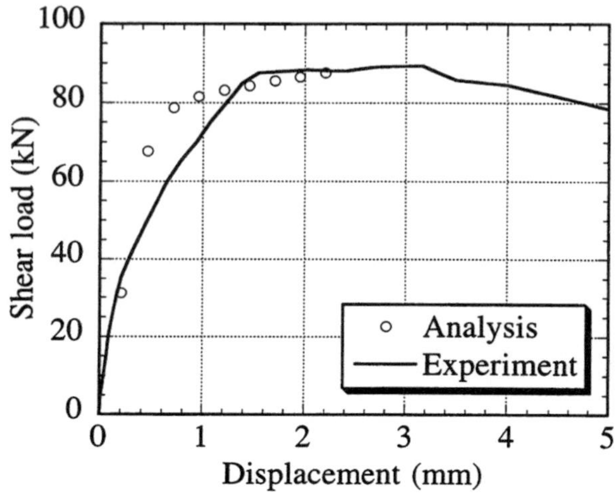

Fig. 2.40 Response of Ohno shear beam calculated by Kanda [12]

2.3.4 *Modified Concrete Models*

In this section we review some other models used for numerical analysis of SHCC, which were obtained by modification of existing models for concrete.

To analyze the behavior of a reinforced ECC (R/ECC) Ohno shear beam, Kanda [12] modified an FE model originally developed for reinforced concrete by Ishida (cited in [12], p. 267). The modification consisted in changing the material uniaxial stress–strain relationship in the crack-normal direction so as to reflect the increased tensile ductility of ECC. The model reproduced the ultimate load carrying capacity under monotonic loading of a shear beam. However, the prepeak load–displacement curve showed higher stiffness – see Fig. 2.40.

Xoxa [11] based his model on the modified compression field theory originally proposed by Vecchio and Collins (cited in [11], p. 110). Xoxa's model variant, which implemented tensile hardening and the assumption that the panel remained a single body even after cracking, captured fairly well the tendency of the prepeak load-deformation curves, but the relative difference of the calculated and experimentally obtained ultimate load ranged from -15% to $+1\%$, depending on the panel reinforcement arrangement.

Suwada and Fukuyama [53] adapted the orthotropic model proposed by Darwin and Pecknold (cited in [53], p. 50) for SHCC materials. The proposed model was based on nonlinear uniaxial stress–strain relations (separate for tension and for compression), which were generalized for a multiaxial stress state by means of failure surface. The tensile behavior of SHCC was represented by a multilinear stress–strain relationship which included both hardening and softening. The model also accounted for reduction of compressive strength due to cracking induced by lateral tension, which was based on experiments. The model was applied to simulate experiments on short reinforced SHCC beam-columns exposed to combination of axial and shear load. The specimens were designed to fail in shear diagonal tension mode (DT series) or in shear compressive mode (SC) series. As seen in Fig. 2.41, the load capacity was captured well by the model in all cases. The hardening response was also well represented for the SC series. However, when the failure was governed by shear-tension (DT series), the numerically reproduced response was much stiffer.

Suryanto et al. [18] modified the model originally developed for reinforced concrete by Maekawa et al. (cited in [18], p. 240) to suit analysis of SHCC elements. The proposed model retained the fundamental features of the original model, including the path-dependent formulations and the assumption of average stress and average strain. The modification mainly consisted of changing the tensile stress–strain relation to represent hardening behavior during multiple cracking phase. In addition, a particular attention was paid to properly treating the shear transfer on cracks, attributed to crack surface friction and fiber bridging. The authors considered that due to the lack of coarse aggregates, cracks in SHCC materials exhibit reduced shear resistance compared to that of ordinary concrete. The proposed model was successfully used to reproduce behavior of shear-critical ECC members exposed to stress field with rotating principal direction (Fig. 2.42, see also Sect. 2.2.2.1).

2.3.5 Individual Crack Based Model

Kabele [54] noticed that constitutive models which treat SHCC in multiple cracking state as an equivalent continuum with monotonous hardening, and thus neglect sequential formation of individual cracks (see Sect. 2.3.3), often exhibit a tendency to overestimate pre-peak stiffness of structural members, in which the material undergoes non-proportional loading. Noting that analyses of such members are often conducted with very fine meshes, in which elements are smaller than the RVE, the author proposes to represent individual cracks in the FE model. Response of such cracks is governed by a traction-separation law, which incorporates bridging effects of both matrix and fibers. As opposed to the earlier discussed homogenized approach, the present traction-separation relation includes a drop of the cohesive traction following matrix cracking and preceding the rebound associated with activation of fiber bridging (Fig. 2.43). Cracks are eventually represented by the smeared approach on an element level, but the constitutive relation is regularized with respect to element size, by which the discrete character of crack treatment is maintained. The model was successfully used to reproduce overall response and shear failure of Ohno beams. More details of these analyses are discussed in Sec. 2.2.2.2.

(P. Kabele)

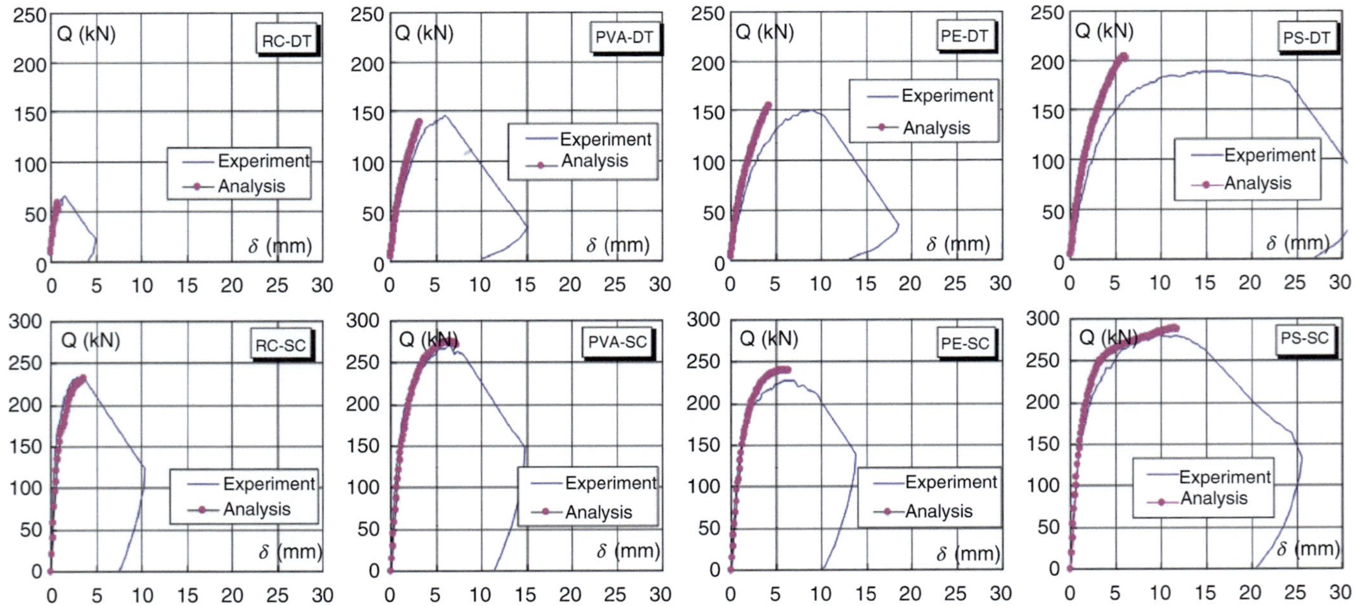

Fig. 2.41 Experimental and analytical results of shear force vs. displacement for short beam-columns exposed to combination of shear and axial compression [53] (RC – ordinary reinforced concrete, PVA, PE, PS – reinforced SHCC)

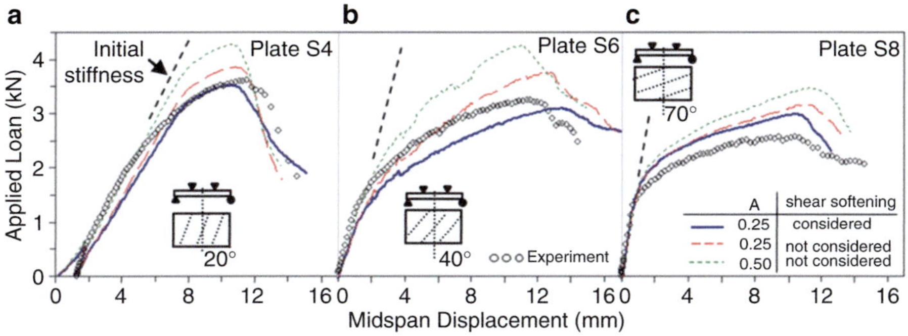

Fig. 2.42 Load-deflection curves of plates with inclined pre-cracks exposed to 4-point bending [18]. Parameter *A* represents the reduction of crack shear resistance

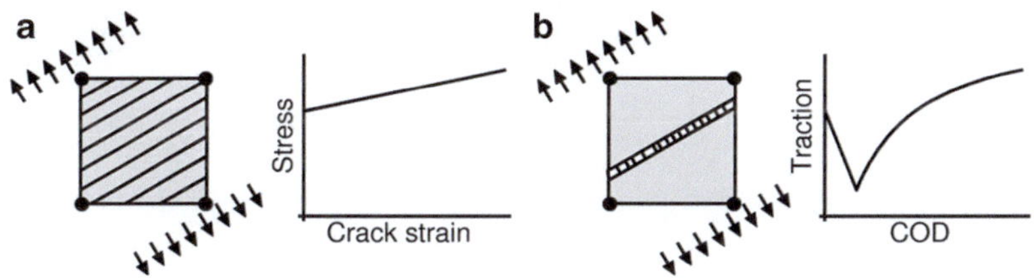

Fig. 2.43 Homogenization based concept (**a**) vs. individual crack based concept (**b**). [54]

2.4 Ductility and Anti-seismic Design

Structural behavior of R/SHCC beams, columns and dampers are compared with those of RC beams columns and dampers, respectively, from the view point of ductility critical to structural elements in aseismic design.

2.4.1 Comparison of Structural Behavior Between R/SHCC Elements and R/C Elements

Structural behaviors of R/SHCC elements are compared with that of R/C elements to indicate the influences of SHCC utilization to the structural performance.

(a) Beams [55]

The Ohno-type cyclic loading test with four beam elements was conducted. The main objective of the test is to investigate the upgrading effects on structural performance in energy dissipation due to ductility and damage tolerance of the R/SHCC beams. PVA fiber with 1.5% in fiber volume fraction was used in the R/SHCC beams. The cross section in clear span of the beams is 200 mm in width, 270 mm in depth, and its length is 1,080 mm. Variables in the test are

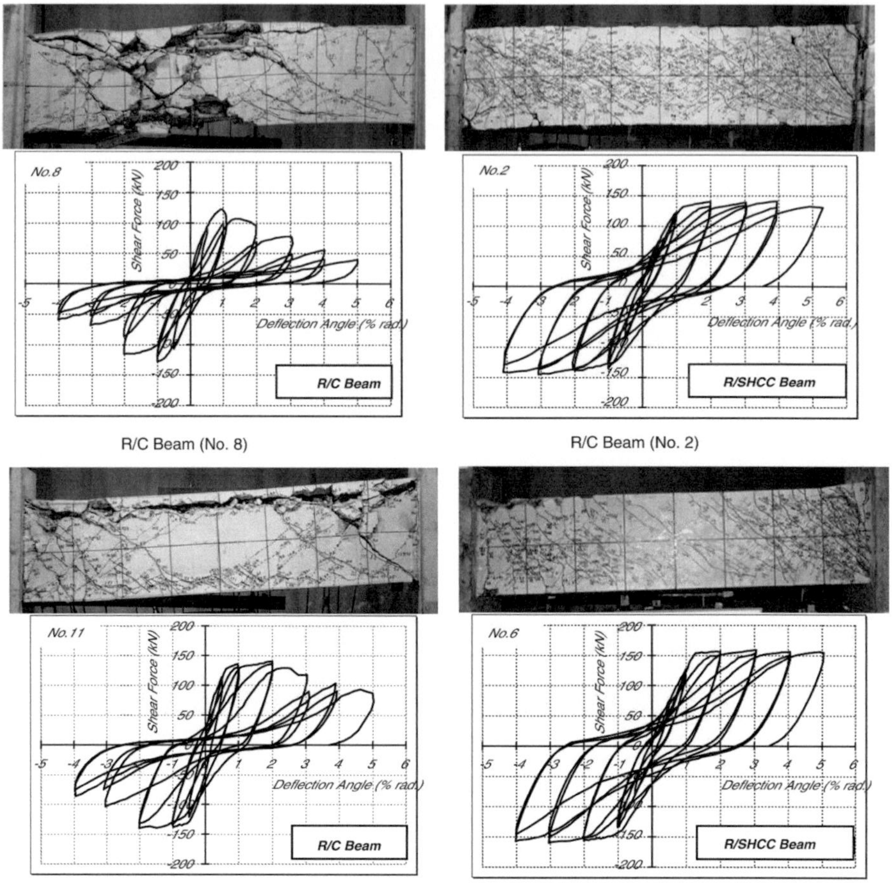

R/C Beam (No. 8) R/C Beam (No. 2)

R/C Beam (No. 11) R/C Beam (No. 6)

Fig. 2.44 Shear force – deflection angle relationships and damage properties at 5% rad

type of cement materials (concrete and SHCC) and bar arrangements due to the planned failure modes, shear failure and bond splitting failure, of the control RC beam specimens.

Figure 2.44 shows comparisons of shear force – deflection angle relationships and damage pattern due to cracking between R/C control beams (No.8 and No.11) and R/SHCC beams (No.2 and No.6). The deflection angle is relative vertical deflection between both ends of the beam divided by its clear span. A pair of RC beam (No.8) and R/SHCC beam (No.6), and a pair of RC beam (No.11) and R/SHCC beam (No.6) have the same bar arrangements, respectively.

The R/C beam (No.8) failed in shear at around 1% in deflection angle as expected in design. On the other hand, R/SHCC beam (No.2) exhibited very ductile structural behavior without shear failure. The damage level due to the crack pattern observed in the R/C beam (No.8) was much higher than that observed in the R/SHCC beam (No.2). Then SHCC material contributed to structural upgrading in shear capacity and damage reduction.

The R/C beam (No.11) failed by bond splitting after yielding at around 2% in deflection angle as expected in design. However, R/SHCC beam (No.6) exhibited very ductile structural behavior with neither bond splitting failure nor bond splitting cracks. The damage level due to the crack pattern observed in the R/C beam (No.11) was much higher than that observed in the R/SHCC beam (No.6). Then SHCC material contributed to structural upgrading in bond splitting capacity and damage reduction.

In both cases, high ductility can be achieved by avoiding any brittle failure like shear failure and bond splitting failure by using SHCC material. Therefore, SHCC material has structural upgrading effects in shear, bond splitting and damage tolerance.

(b) Columns [55]

The cyclic loading test under anti-symmetrical moment condition (double moment condition) of two column elements was conducted. The objective of the column test is the same as that of the beam test.

The configuration and bar arrangements are shown in Fig. 2.45a. The cross section in clear span of the columns is 300 mm square and its length is 900 mm. The axial force applied to the column is 20% of the axial compressive strength of the column, which calculated ignoring the contribution of the steel bars.

Damage properties and shear force – deflection angle relationships of both specimens are illustrated in Fig. 2.45c. The damaging process and failure mode of the RC column was similar to those observed in the R/C beam (No.8). However large bond splitting cracks were observed as shown in Fig. 2.45b. The R/C column failed in shear without yielding of the longitudinal bars. After that, the resistant shear force in the skeleton curve of shear force – deflection angle relationship decreased with increase of deflection angle. On the other hand, the R/SHCC column failed in neither shear nor bond splitting. It showed very good ductile behavior up to the end of the test with initiation of the fine cracks as shown in the beam tests. Since the damage of the R/SHCC column was also much lighter than that of the R/C column, SHCC could upgrade the damage tolerance of seismic resistant structural elements.

(c) Dampers (Ductile elements with small depth to length ratio) [13]

The structural performance of R/SHCC damper with small depth to length ratio used for response control of RC structures was examined using the static loading test. The test results of a R/SHCC damper were compared with those of mortar damper reinforced with steel hoops (R/Mortar) as shown in Fig. 2.46. R/Mortar were badly damaged and eventually fractured due to the large shear force and compressive force. The deformation capacity of R/Mortar was less than 3%. On the other hand, large deflection capacity of over 13% rad. under the very high average shear stress of 6 N/mm^2 and damage reduction effect were observed in case of the R/SHCC damper which has the same bar arrangements as the case of R/mortar.

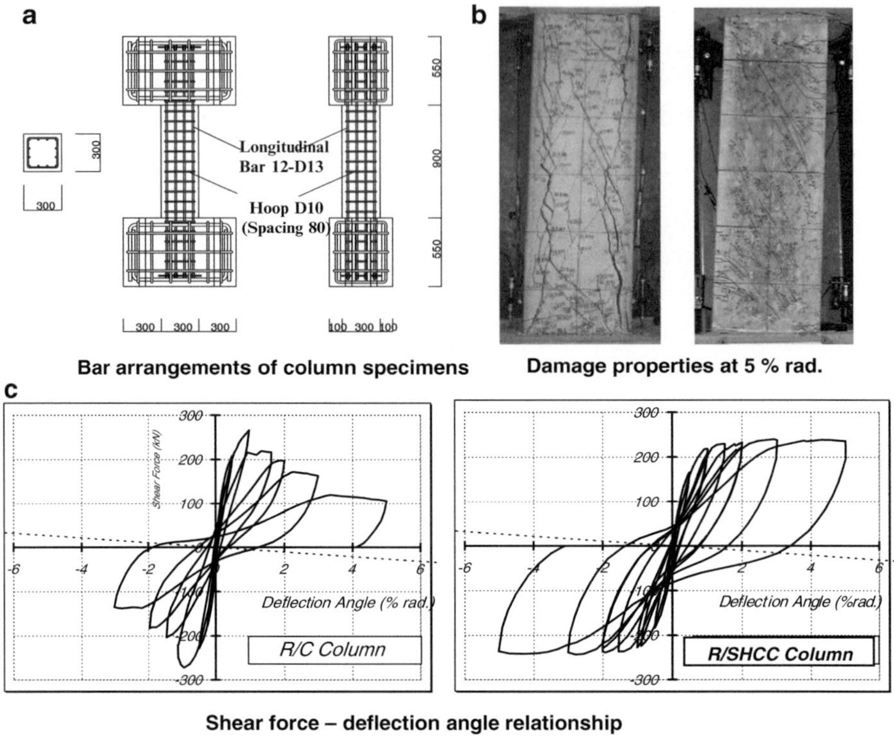

Fig. 2.45 Bar arrangements and test results of column specimens

As shown in Figs. 2.44, 2.45 and 2.46, SHCC materials have a reinforcing effect for shear and bond splitting. Thus, very ductile structural behavior can be achieved by preventing brittle failures, shear failure and bond splitting failure by using SHCC instead of concrete.

2.4.2 Design Concept of RC Elements and Advantage of SHCC Utilization

Capacity oriented design or ductility oriented design are generally applied for seismic design of R/C structures.

Sufficient amount of strong structural elements like R/C shear walls are arranged to resist against big earthquake motion in the capacity oriented design. The maximum capacity of structural elements when brittle failure occur can be drastically increased by using SHCC instead of concrete because of the reinforcing effect of SHCC. Usually, maximum shear stresses of R/SHCC elements are much higher than those of R/C elements. The increase of elemental capacity leads to increase of structural capacity.

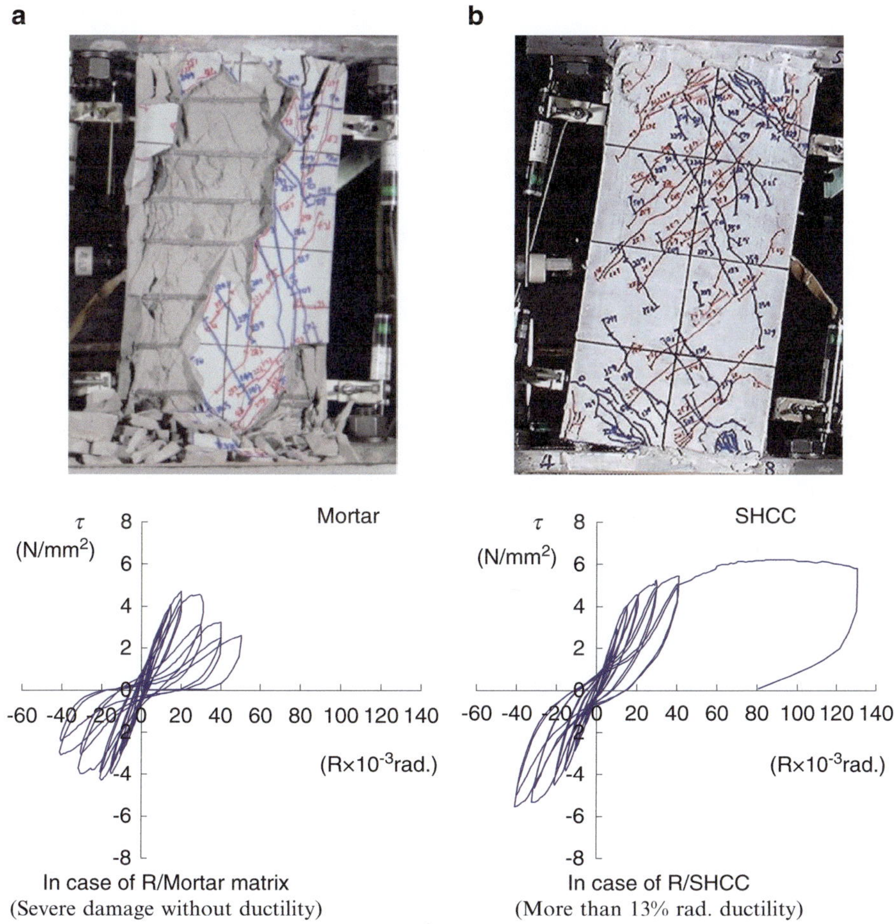

Fig. 2.46 Lateral-loading-test results of HPFRCC device compared with mortar device

On the other hand, seismic energy can be dissipated due to ductility of structural elements by avoiding any brittle failure in ductility oriented structural design. Elemental configuration, bar arrangements, strength of materials and so on should be appropriately adjusted to obtain the necessary ductility of the elements. The integrity concrete and steel reinforcement in R/C elements is easily collapsed when the elements deflect largely. Then shear failure or bond splitting failure after yielding of elements may occur in case that the ratio of shear capacity (or bond splitting capacity) to flexural capacity decreases toward 1. However, since SHCC can maintain integrity between steel reinforcement and SHCC at large deflection, R/SHCC elements exhibit very ductile structural behavior avoiding any brittle failure.

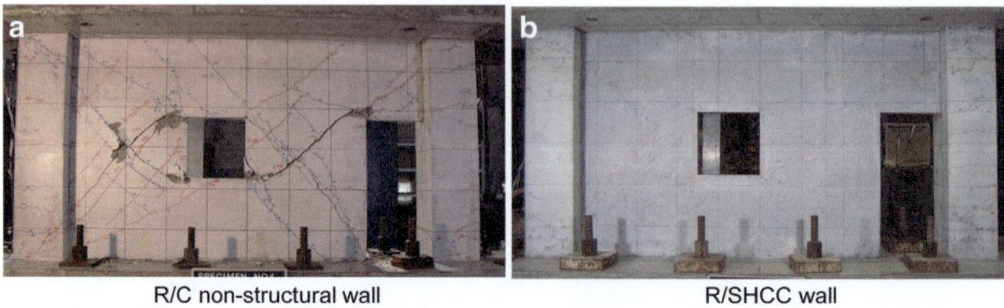

Fig. 2.47 Comparison of damage at 1/200 rad. of non-structural walls

Suwada [36] proposed an evaluation method for ductility of R/SHCC elements using shear capacity to flexural capacity ratio. This ratio is already used for R/C buildings in seismic retrofitting design in Japan [56]. However, there are not enough test data to evaluate the ductility of structural elements with SHCC materials.

Fukuyama et al. [57] proposed an application method of SHCC to non-structural walls whose contribution to seismic resistance of structures is ignored in the usual structural design. Since the capacity and ductility of the non-structural wall with SHCC can be drastically increased, its structural contribution can be counted in the seismic design. It is not non-structural wall any more. However, evaluation method to predict the capacity and ductility of new R/SHCC elements should be developed for taking into account its contribution in the design.

The seismic design of structures with R/SHCC elements can be performed using the capacity spectrum method or time history earthquake response analysis if we have information of necessary data for these analyses like lateral force – deflection relationships (Fig. 2.47).

(H. Fukuyama and H. Suwada)

References

1. Japan Society of Civil Engineers: Recommendations for Design and Construction of High Performance Fiber Reinforced Cement Composite with Multiple Fine Cracks. Concrete Library 127/Japan Society of Civil Engineers, Tokyo (2007)
2. Hou, W., Shi, Q.X., Ma, Z.L.: Seismic behavior analysis on reinforced concrete core walls based on fiber model. J. Adv. Mater. Res. **163–167**, 1068–1073 (2010)
3. Uchida, Y., Lim, S, Moriyama, M.: Cracking behavior of reinforced concrete beams repaired with HPFRCC. In: Proceedings of the 4th International Conference on Construction Materials: Performance, Innovations and Structural Implications, Nagoya, Japan, pp. 327–323 (2009)
4. Kanakubo, T., Shimizu, K., Kanda, T.: Size effect on flexural and shear behavior of PVA-ECC structures under extreme loading. Structures under Extreme Loading. Proceedings of Protect 2007 Workshop, Paper No. 54 (2007)
5. JCI Technical Committee: Method of test for bending moment-curvature curve of fiber-reinforced cementitious composites (JCI-S-003-2005). J. Adv. Concr. Technol. **4**, 73–78 (2006)

6. Lepech., M., Li, V.C.: Preliminary findings on size effect in ECC structural members in flexure. In: Proceedings of the International Symposium on Brittle Matrix Composites 7, Warsaw, ZTUREK RSI and Woodhead Publications, pp. 57–66 (2003)
7. Canbolat, B.A., Parra-Montesinos, G., Wight, JK.: Effect of fiber reinforced cementitious composites on the seismic behavior of RC coupling beams. In: Proceedings of the fib Symposium, Concrete Structures in Seismic Regions, Athens, May 2003
8. Nagai, S., Kanda, T., Maruta, M., Miyashita, T.: Shear capacity of ductile wall with high performance fiber reinforced cement composite. In: Proceedings of the 1st fib Congress, Osaka, Japan pp. 767–774 (2002)
9. Nagai, S., Kaneko, T., Kanda, T., Maruta, M.: Structural capacity of reinforced PVA-ECC dampers. In: Proceedings of the BEFIB 2004, Varenna, Lake Como, Italy (RILEM Publications S.A.R.L., 1999, 2004) pp. 1227–1236 (2004)
10. Shimizu, K., Kanakubo, T., Kanda, T., Nagai, A.: Shear behavior of steel reinforced PVA-ECC beams. In: Proceedings of the 13th World Conference on Earthquake Engineering, in DVD paper no. 704, Vancouver, BC (2004)
11. Xoxa, V.: Investigating the shear characteristics of high performance fiber reinforced concrete. M.Sc. thesis, University of Toronto (2003)
12. Kanda, T.: Design of engineered cementitious composites for ductile seismic resistant elements. Ph.D. thesis, University of Michigan (1998)
13. Fukuyama, H., Suwada, H.: Experimental response of HPFRCC dampers for structural control. J. Adv. Concr. Technol. **1**, 317–326 (2003)
14. Kesner, K.E.: Development of seismic strengthening and retrofit strategies for critical facilities using engineered cementitious composite materials. Ph.D. thesis, Cornell University (2003)
15. Kesner, K.E., Billington, S.L.: Experimental response of precast infill panel connections and panels made with DFRCC. J. Adv. Concr. Technol. **1**, 1–7 (2003)
16. Chao, S.H., Naaman, A.E., Parra-Montesinos, G.J.: Bond behavior of strand embedded in fiber reinforced cementitious composites. PCI J. **51**, 56–71 (2006)
17. Kanakubo, T., Shimizu, K., Kanda, T., Nagai, S.: Shear transmission on crack surface of ECC. In: Proceedings 7th International Conference on Fracture Mechanics of Concrete and Concrete Structures (FraMCoS-7), Jeju, Korea (2010)
18. Suryanto, B., Nagai, K., Maekawa, K.: Modeling and analysis of shear-critical ECC members with anisotropic stress and strain fields. J. Adv. Concr. Technol. **8**, 239–258 (2010)
19. Suryanto, B., Nagai, K., Maekawa, K.: Bidirectional multiple cracking tests of HPFRCC plates. ACI Mater. J. **107**, 450–460 (2010)
20. Fukuyama, H., Matsuzaki, Y., Nakano, K., Sato, Y.: Structural performance of beam elements with PVA-ECC. In: Proceedings HPFRCC3, Mainz, Germany (RILEM Publications S.A.R.L.), pp. 531–541 (1999)
21. Li, V.C., Mishra, D.K., Naaman, A.E., Wigh, J.K., LaFave, J.M., Wu, H.C., Inada, Y.: On the shear behavior of engineered cementitious composites. J. Adv. Cem. Based Mater. **1**, 142–149 (1994)
22. Kanda, T., Watanabe, S., Li, V.C.: Application of pseudo strain hardening cementitious composites to shear resistant structural elements. In: Proceedings FRAMCOS-3, Fracture Mechanics of Concrete Structures, AEDIFICATIO Publishers, Freiburg, Germany, pp. 1477–1490 (1998)
23. Shimizu, K., Kanakubo, T., Kanda, T., Nagai, S.: Shear behavior of PVA-ECC Beams. In: Proceedings of the International Workshop on High Performance Fiber Reinforced Cementitious Composites in Structural Applications, Honolulu, Hawaii (2005)
24. Kim, J.H., Lim, Y.M., Won, J.P.: Experimental investigation of shear dominant RC beam repaired with DFRCC at Cover Thickness. In: Proceedings of the International Workshop on High Performance Fiber Reinforced Cementitious Composites in Structural Applications, Honolulu, Hawaii (2005)
25. Canbolat, B.A., Parra-Montesinos, G.J., Wight, J.K.: Experimental study on seismic behavior of high-performance fiber-reinforced cement composite coupling beams. ACI Struct. J. **102**, 159–166 (2005)

26. Parra-Montesinos, G.J., Chompreda, P.: Deformation capacity and shear strength of fiber-reinforced cement composite flexural members subjected to displacement reversals. J. Struct. Eng. **133**, 421–431 (2007)
27. Fukuyama, H., Suwada, H., Mukai, T.: Test on high-performance wall elements with HPFRCC. In: Proceedings of the International Workshop on High Performance Fiber Reinforced Cementitious Composites in Structural Applications, Honolulu, Hawaii (2005)
28. Kim, K., Parra-Montesinos, G.J.: Behavior of HPFRCC low-rise walls subjected to displacement reversals. In: Naaman, A.E., Reinhardt, H.W. (eds.) Fourth International RILEM Workshop on High Performance Fiber Reinforced Cement Composites (HPFRCC 4), Ann Arbor, USA, RILEM Publications s.a.r.l. pp. 505–515 (2003)
29. Parra-Montesinos, G.J., Wight, J.K.: Seismic response of exterior RC column-to-steel beam connections. ASCE J. Struct. Eng. **126**, 1113–1121 (2000)
30. Parra-Montesinos, G.J., Peterfreund, S.W., Chao, S.H.: Highly damage-tolerant beam-column joints through use of high-performance fiber-reinforced cement composites. ACI Struct. J. **102**, 487–495 (2005)
31. Kabele, P.: Finite element fracture analysis of reinforced SHCC members. In: Proceedings of the Advanced Concrete Materials 2009, Stellenbosch, South Africa (2009)
32. Architectural Institute of Japan: Design Guidelines for Earthquake Resistant Reinforced Concrete Buildings Based on Ultimate Strength Concept. AIJ (in Japanese), Tokyo (1990)
33. Nagai, S., Kaneko, T., Kanda, T., Maruta, M.: Structural capacity of dampers with PVA-ECC. Proc. Jpn. Concr. Inst. **26**, 1513–1518 (2004) (in Japanese)
34. Shimizu, K., Kanakubo, T., Kanda, T., Nagai, A.: Evaluation of shear strength for PVA-ECC member and shear transfer mechanism on crack surface. J. Struct. Constr. Eng. (Trans. AIJ) **619**, 133–139 (2007) (in Japanese)
35. Japan Society of Civil Engineers: Recommendations for Design and Construction of High Performance Fiber Reinforced Cement Composites with Multiple Fine Cracks (Hpfrcc). Japan Society of Civil Engineers/Concrete Library 127, Tokyo (2008)
36. Suwada, H.: Study on shear behavior of dampers using high performance fiber reinforced cement composites with strain hardening type tensile properties. Ph.D. thesis, Chiba University (in Japanese) (2008)
37. Dick-Nielsen, L.: Modeling of ECC materials using numerical formulations based on plasticity. Ph.D. thesis, Technical University of Denmark (2008)
38. Dick-Nielsen, L., Stang, H., Poulsen, P.N. Condition for strain-hardening in ECC uniaxial test specimen. In: Konsta-Gdoutos, M.S. (ed.) Proceeding of the Measuring, Monitoring and Modeling Concrete Properties, Evanston, USA, pp. 41–47 (2006)
39. Li, V.C., Fischer, G.: Reinforced ECC – an evolution from materials to structures. In: Proceedings of the First fib 2002 Congress, Osaka, Japan (2002)
40. Kabele, P.: Analytical modeling and fracture analysis of engineered cementitious composites. Ph.D. thesis, The University of Tokyo (1995)
41. Kabele, P., Horii, H.: Analytical model for fracture behavior of pseudo strain-hardening cementitious composites. Concr. Libr. Int. **29**, 105–120 (1997)
42. Kullaa, J.: Finite element modelling of fibre-reinforced brittle materials. Heron **42**, 75–97 (1997)
43. Kabele, P., Takeuchi, S., Inaba, K., Horii, H.: Performance of engineered cementitious composites in repair and retrofit: analytical estimates. In: Reinhardt, H.W., Naaman, A. (eds.) High Performance Fiber Reinforced Composites – Proceeding HPFRCC3, pp. 617–627 (1999)
44. Boshoff, W.P., van Zijl, G.P.A.G.: A computational model for strain-hardening fibre-reinforced cement-based composites. J. S. Afr. Inst. Civ. Eng. **49**, 24–31 (2007)
45. van Zijl, G.P.A.G., Boshoff, W.P.: Mechanisms of creep in fibre-reinforced Strain-Hardening Cement Composites (SHCC). In: 8th International Conference on Creep, Shrinkage and Durability Mechanics of Concrete and Concrete Structure, Shima Kanko Hotel, Ise-Shima, Japan, 30 September 2008 through 2 October 2008, CRC Press, Boca Raton, pp. 753–759 (2008)

46. Takeuchi, S.: Modeling of ECC fracture behavior under shear and analysis of its performance in antiseismic retrofit. Master's thesis, University of Tokyo (1999)
47. MARC Analysis Research Corporation.: MARC Volume D: User Subroutines and Special Routines K7 (1997)
48. Inaba, K.: Analytical study on the repair of concrete structures with ECC. Master's thesis, University of Tokyo (1999)
49. Kabele, P.: Equivalent continuum model of multiple cracking. Engineering mechanics. Assoc. Eng. Mech. Czech Repub **9**, 75–90 (2002)
50. Kabele, P.: Assessment of structural performance of engineered cementitious composites by computer simulation. Habilitation thesis, CTU in Prague, CTU Reports 4(5) (2001)
51. Han, T.S., Feenstra, P., Billington, S.L.: Simulation of highly ductile fiber-reinforced cement-based composite components under cyclic loading. ACI Struct. J. **10**, 749–757 (2003)
52. Kesner, K., Billington, S.: HPFRCC infill panels under cyclic loading: experiments and numerical simulations. In: Fischer, G., Li, V.C. (eds.) Proceedings of the International RILEM Workshop on High Performance Fiber Reinforced Cementitious Composites (HPFRCC) in Structural Applications, Honolulu, USA, pp. 393–402 (2006)
53. Suwada, H., Fukuyama, H.: Nonlinear finite element analysis of shear failure of structural elements using high performance fiber reinforced cement composite. J. Adv. Concr. Technol. **4**, 45–57 (2006)
54. Kabele, P.: Finite element fracture analysis of reinforced SHCC members. In: van Zijl, G.P.A.G., and Boshoff, W.P. (eds.) Advances in Cement-Based Materials. Stellenbosch, South Africa: CRC Press, Taylor & Francis Group. pp. 237–244 (2009)
55. Fukuyama, F., Matsuzaki, Y., Sato, Y., Iso, M., Suwada, H.: Structural performance of engineered cementitious composite elements. In: Xiao, Y., Mahin, S. (eds.) Composite and Hybrid, Proceedings of the 6th ASCCS (Association for International Cooperation and Research on Steel-Concrete Composite Structures) Conference, Los Angeles, CA, pp. 969–976 (2000)
56. Japan Building Disaster Prevention Association.: Guidelines for Seismic Evaluation of Existing Reinforced Concrete Buildings (2005)
57. Fukuyama, H., Iso, M., Ogawa, A., Suwada, H.: Mitigation of damage due to crack of RC elements utilizing high performance fiber reinforced cementitious composites. In: Reinhardt, H.W., Naaman, A.E. (eds.) Proceedings of the Fifth International RILEM Workshop on High Performance Fiber Reinforced Cement Composites (HPFRCCS-5), RILEM Proceedings PRO53, RILEM Publications S.A.R.L., Cachan, France, pp. 427–435 (2007)
58. Kabele, P., Kanakubo, T.: Experimental and numerical investigation of shear behavior of PVA-ECC in structural elements. In: Reinhardt, H.W., Naaman, A. (eds.) Proceedings of the Fifth International Workshop on High Performance Fiber Reinforced Cementitious Composites (HPFRCC5), Mainz, Germany (2007)

Chapter 3
Application of SHCC

Abstract Utilizing concept for SHCC as a new material, SHCC material evaluation for application, and actual application examples are presented in this chapter. SHCC has been applied to newly constructed structures, making use of its excellent mechanical performance. It has also been used for surface repair of existing concrete structures making use of its finely distributed cracking behavior. Appropriate use of the tensile performance can work out a structural component excellent in both durability and mechanical performance.

Keywords Structural element • Anti-seismic • Repair • Retrofit • Fatigue • High-rise building • Steel bridge

3.1 Introduction

SHCC shows remarkably high ductility under uniaxial tensile stress and excellent performance distinguished from conventional cementitious materials by multiple cracking and strain hardening behavior. Characteristics of SHCC include crack width controlling capability keeping crack width in a permissible range. SHCC has been applied to bridge decks and building dampers, making use of its excellent mechanical performance. It has also been used for surface repair of concrete dams, water channels, and retaining walls making use of its finely distributed cracking behavior. Appropriate use of the tensile performance can work out a structural component excellent in both durability and mechanical performance.

A technical committee in Japan Concrete Institute, Performance Evaluation and Structural Use of Highly Ductile Cementitous Composite (Chairman Prof. Rokugo, from 2001 to 2004), played an important role in the progress of application in SHCC. Inheritance of the technical committee was developed in two JSCE committees, the subcommittee of the Concrete Committee for Evaluation and Use of the Fiber Reinforced Mortars with Multiple Fine Cracks (Chairman Prof. Rokugo, from 2004 to present) and the Drafting of the Guideline for Multiple Crack Type

K. Rokugo and T. Kanda (eds.), *Strain Hardening Cement Composites: Structural Design and Performance*, RILEM State-of-the-Art Reports 6, DOI 10.1007/978-94-007-4836-1_3, © RILEM 2013

Fiber Reinforced Cement Composite (Chairman Prof. Rokugo, from 2005 to 2008) which led to the publication of world's first guideline for design and construction of SHCC. This may result in a milestone for a broad dissemination of the material and an evidence of foregoing Japanese research on SHCC.

In the following, utilizing concept for new material like SHCC, SHCC material evaluation for application, and actual application examples are presented.

(T. Kanda)

3.2 A Concept for Use of New Materials

3.2.1 Performance Based Design for SHCC Application

Performance-based design is a sophisticated design concept that allows engineers to set freely the level of structural performance. Under this concept, the structural designer creates the building structure based on the required structural performance, evaluates whether it is satisfied, and the evaluated performance is stated.

When structural performance is evaluated and the results are disclosed, clients will be able to regard building structures as market products whose performance level and total cost can be compared just like automobiles and computers (Fig. 3.1). Clients will then be able to request an appropriate level of performance, which is not necessarily the minimum level requirement stated in the present code.

The social requirements for infrastructure and building structures have been ever changing along with social and economic development. As peoples' lifestyles and social activities have diversified in recent years, the performance items and levels that people require to infrastructure and building structures have become more diversified too. This tendency will continue in the future as long as society keeps maturing. Technical development in the twenty first century should aim to adequately meet such changing requirements of an evolving society. Therefore, the development of new technologies and materials to realize high performance that meets social requirements adequately will be expected. The following concepts are indispensable for utilization of new materials.

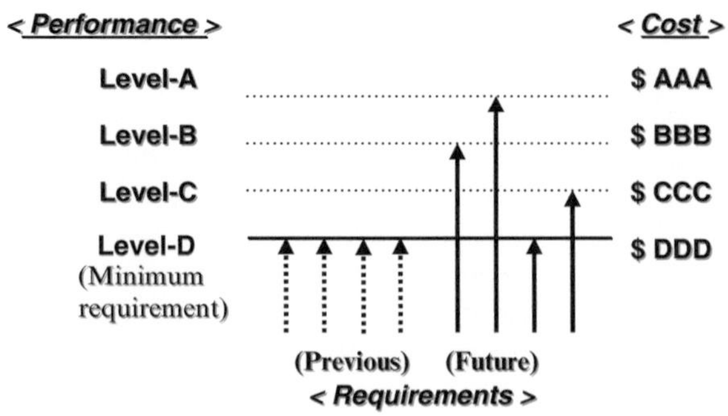

Fig. 3.1 Performance menu (comparison between performance level and cost)

(i) New materials should coexistence with conventional materials
Conventional social requirements have been met by using conventional materials appropriately. Then, new materials are not necessary to use for this objective. The conventional materials can continue to be used. The target area for applying the new materials is the elements that the new types of performance or higher level in conventional performance item are required. Then new materials should coexist with conventional materials. However, their role should be different.

(ii) Watch the social change
Performance requirement is always based on the social and economical condition, any social and/or economical changes should not be missed. Then it is very important to watch any sign of change related to performance requirements of structures.

(iii) Cooperation between material engineers and structural engineers including structural designers is indispensable
The strong cooperation between specialists in the material engineering area and specialists in the structural engineering area, especially structural designer, is highly expected. It is indispensable to seek new technology using new materials in order to meet the new social requirements. Close discussion between people from material engineering and structural engineering should be accelerated.

3.2.2 *Potential Application*

Following are examples of ideas of SHCC application based on "a concept for use of new materials" described in Sect. 3.1. Experimental and analytical research were conducted to investigate structural performance of the structures or members.

3.2.2.1 SHCC Dampers [1]

One of the lessons from the 1995 Kobe earthquake is many buildings with soft first story damaged severely even though the buildings were designed according to the current code. Then it is necessary to develop the strengthening method for the buildings with soft first story without reducing open space in the first story. Fukuyama [1] introduced an application example of SHCC dampers used for increasing structural safety by reducing response drift of the first story as well as increasing structural resiliency by reducing structural damage. The SHCC damper is a SHCC short column reinforced with steel bars that has very high strength, stiffness and ductility compared with conventional RC columns with the same configuration and bar arrangement. An analytical study on the seismic response of a soft first story building with and without such SHCC dampers was performed as a case study to

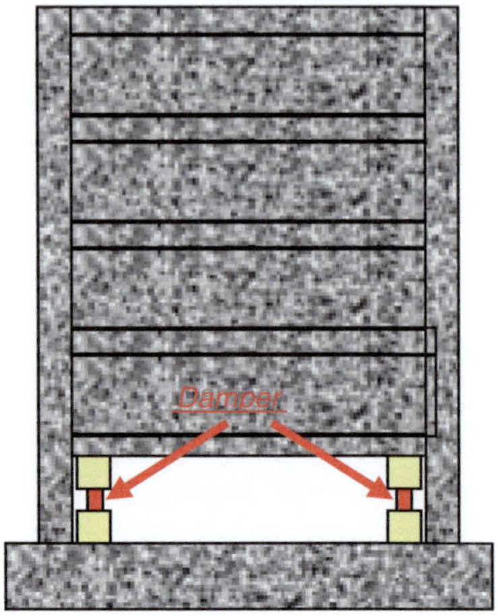

Fig. 3.2 SHCC dampers for response control of buildings with soft first story

investigate the feasibility of the proposed technique for damage mitigation against large earthquakes. The results indicate that SHCC dampers can reduce the drift angle of the soft first story from 2% to 0.5% in the case of seismic input with maximum velocity normalized at 50 cm/s. Since a drift angle of 0.5% means an elastic response of the structure, SHCC devices are confirmed to have significant potential as a new structural technology for structural safety and damage mitigation (Fig. 3.2).

3.2.2.2 Ductile Shear Wall [2]

RC Shear wall is an effective element for seismic retrofit to increase only capacity of RC buildings. However, if they have large ductility to be a ductile shear wall, the seismic retrofit with them will be much more effective to increase structural safety by increasing capacity and ductility. Fukuyama et al. [2] introduced one type of ductile shear wall which is consisted with a couple of SHCC dampers and experimentally investigated that this wall exhibited high strength and high ductility (Fig. 3.3).

3.2.2.3 SHCC Wall with Large Opening [3]

RC walls can be classified into two categories. One is a shear wall with thickness of 12 cm or larger and with enough amount of steel bar in vertical and horizontal exhibits large lateral capacity. Another one is non-structural wall with large openings in the usual case whose contribution to seismic resistance of structures is ignored in the usual structural design. This is because the stiffness and lateral capacity of the RC non-structural wall are much smaller than those of shear wall.

However, Fukuyama et al. [3] introduced that if the SHCC is used instead of concrete of non-structural wall, its capacity and ductility can be drastically

Fig. 3.3 "Ductile shear wall" with multiple SHCC dampers

increased. Then the structural contribution such as capacity and damage tolerance of SHCC wall with large opening can be expected in the seismic design. Figure 2.39 shows the contrast of damage states at 1/200 rad in drift angle. These facts demonstrate structural wall with large opening, which is a new type of structural elements, can be established by using SHCC.

(H. Fukuyama and H. Suwada)

3.3 SHCC's Material Characterization for Application

A unique feature of SHCC lies in tensile performance including pseudo strain-hardening tensile stress–strain behavior and crack width controlling capability, while compressive performance is not very different from that of the normal concrete. To reflect the tensile performance in structural design, three major material properties, tensile yield strength, ultimate tensile strain capacity and maximum crack width, are defined. The first two characteristics can be determined by a direct tensile test [4] as shown in Fig. 3.4. When the tensile stress–strain relation is determined with the proposed method, sample data of the tensile yield strength and the ultimate tensile strain capacity can be obtained as shown in Fig. 3.5, and the characteristic values of tensile yield strength f_{tyk} and ultimate tensile strain capacity ε_{tuk} are to be determined taking account of their variations. Examples of variation in the tensile yield strength and the ultimate tensile strain capacity are shown in Figs. 3.6 and 3.7. A schematic representation of the characteristic values and the measured sample data is shown in Fig. 3.8.

The characteristic value of crack width can be defined as the maximum crack width for which a test method is proposed [5]. In this test method, crack widths as shown in Fig. 3.9 are measured under the tensile loading performed in the same way as in Fig. 3.4. The crack width measurement can be executed either with direct

Fig. 3.4 Direct uniaxial tensile test

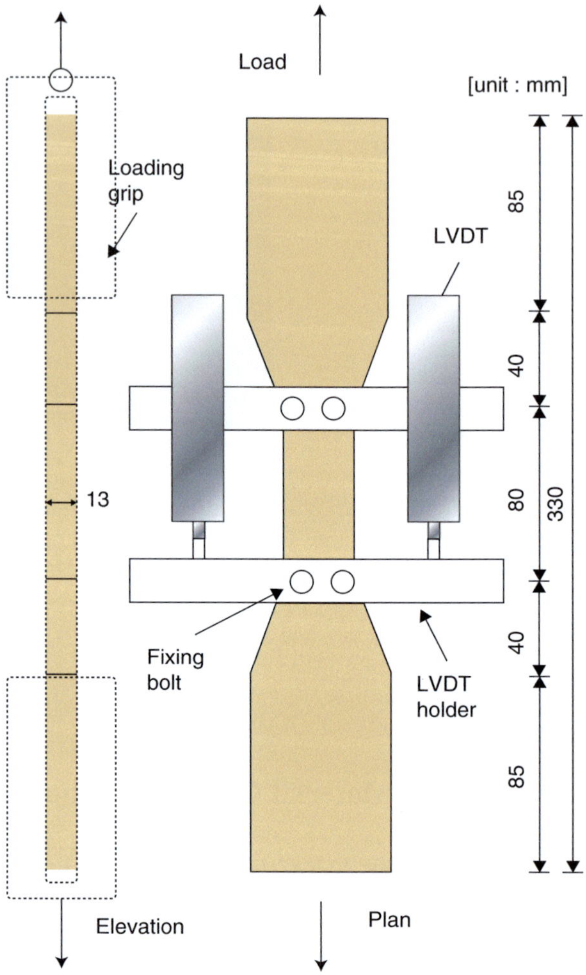

Fig. 3.5 Method of tensile characteristics evaluation

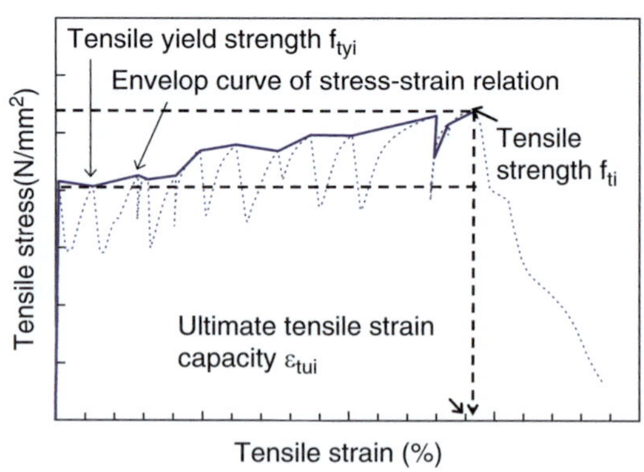

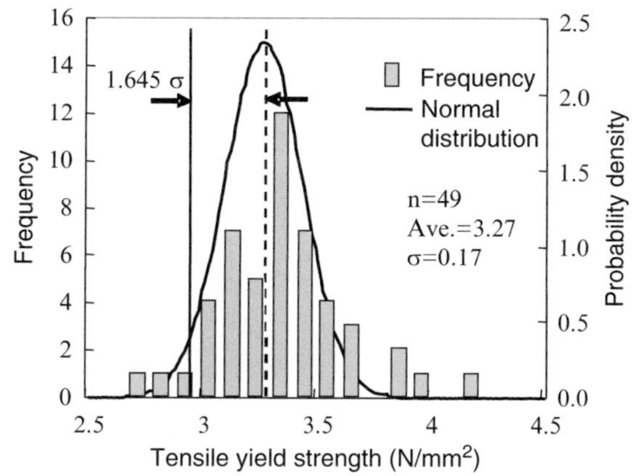

Fig. 3.6 Distribution of measured tensile yield strengths

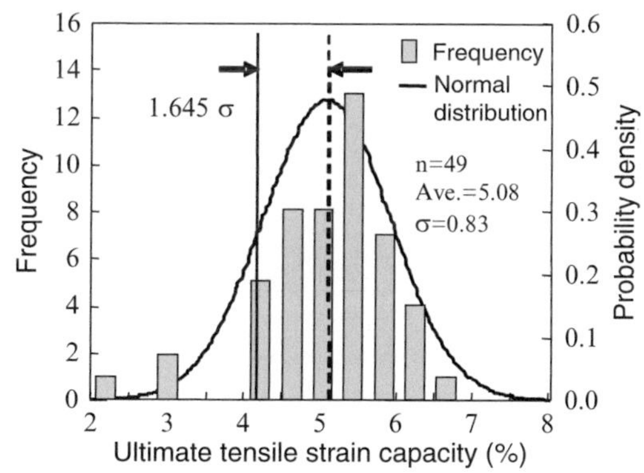

Fig. 3.7 Distribution of measured ultimate strains

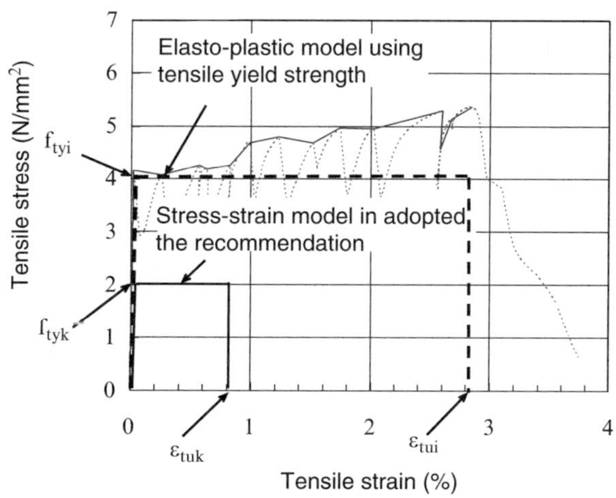

Fig. 3.8 Measured tensile stress

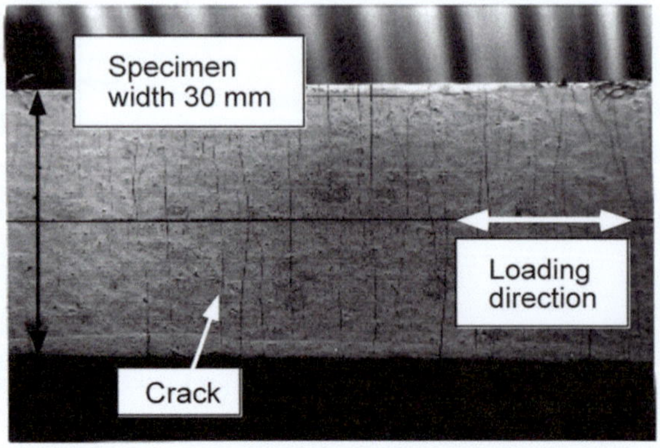

Fig. 3.9 Multiple fine cracks in ECC and characteristic values

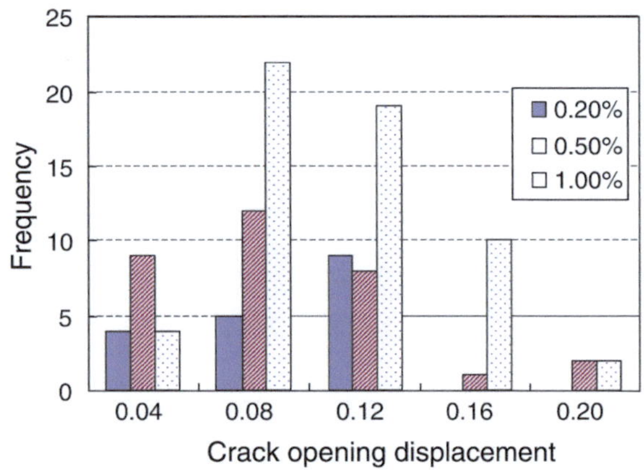

Fig. 3.10 Distribution of measured crack widths at different tensile strain level

microscope observation or indirect estimation based on the number of cracks and the tensile strain. When the crack width data are obtained, the maximum crack width is determined taking account of variations as shown in Fig. 3.10.

(T. Kanda)

3.4 Application Examples

3.4.1 Advantages for R/SHCC Application

Among SHCC materials, Engineered Cementitious Composite – ECC shows excellent material properties and applicability and a wider dissemination is expected [6]. Hereafter we will mainly deal with ECC regarding the state-of-the-art of research, problems and future prospects.

A structural member combined with steel members and ECC is called R/ECC (abbreviation of Reinforced ECC) and shows unique characteristics largely different from the conventional RC and PC, which is regarded as a structural form of new generation. As well known, structural form of both RC and PC expect compressive resistance from concrete, tensile resistance from steel and shear resistance both from concrete and steel placing right materials in the right positions. While in R/ECC where concrete is replaced with ECC, the role of concrete for shear resistance is extended over tensile resistance and an increase in tensile strength by addition of ECC to steel can be expected.

R/ECC also shows an interesting feature in protecting steel from corrosion unlike other structural forms. Cracking at working conditions is not perfectly avoidable in concrete members and ingress of harmful species for corrosion through the cracking parts may lead to a progress of steel reinforcement corrosion resulting in degradation of structural members. However in R/ECC, it was proven that narrow crack widths and generation of multiple cracks prevent mass transport resulting in a lower corrosion rate than that of normal concrete. Hence R/ECC has excellent corrosion protection capability for reinforcing steels and can restrict structural members from degradation due to steel corrosion.

Examples of applying these characteristics, tensile resistance and corrosion protection capability, will be introduced in the following chapters. Full use of R/ECC as a third generation of structural form other than RC and PC could realize a structure with excellent durability and safety in a cost-effective manner.

3.4.2 Use of Tensile Performance for Tunnel Lining

As an example using tensile resistance of ECC, application to inner lining of tunnel is first introduced here. There are several options of reinforcing measure for considerably damaged tunnels but all of them are expensive and are likely to take longer construction duration. Thick inner reinforcing construction may often invade inner space for service. Hence thin inner reinforcing construction method with compressive and tensile resistance, safety and cost effectiveness is needed. An inner reinforcing construction using R/ECC has been proposed to meet the requirement [4]. In this construction method, expanded metals are fixed on the inner wall surface on which ECC is sprayed. The resulting thickness of 50 to 70 mm is a major characteristic and almost a half of the conventional constructions as shown in Fig. 3.11. Further advantages include excellent tensile resistance of ECC and steels to local stresses that occur in the tunnel lining, and minimization of flaking and de-lamination risks by fixing steels on the existing liner before ECC spraying.

A full-scale loading test for the R/ECC liner construction as shown in Fig. 3.12 was reported [7]. It was confirmed that the maximum load after R/ECC construction was 1.3 times greater than that before construction (Fig. 3.13).

This method was applied to the emergency recovery construction of the Tennou tunnel of JR Joetsu line damaged by the Niigata Chuetsu Earthquake [4].

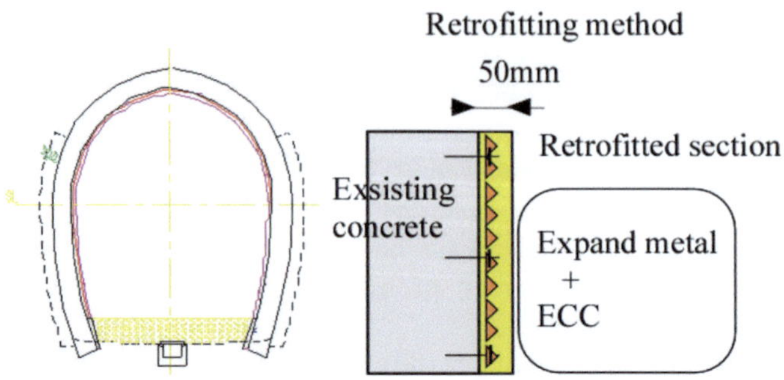

Fig. 3.11 Tunnel liner reinforcing construction with R/ECC

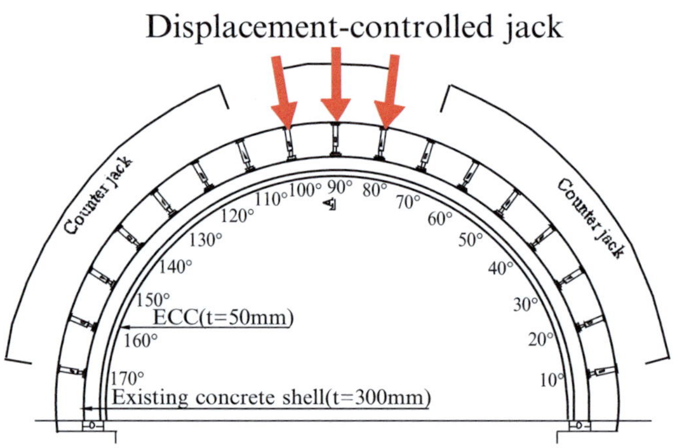

Fig. 3.12 Loading test of R/ECC liner reinforcing construction

Fig. 3.13 Results of the loading test of R/ECC liner reinforcing construction

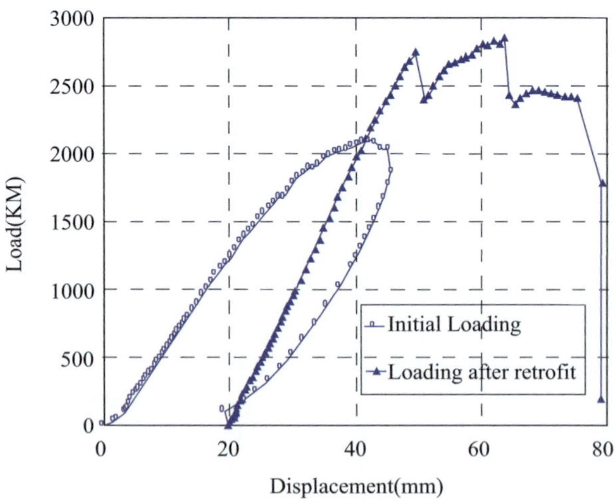

Fig. 3.14 Execution of R/ECC liner reinforcing construction

The Ten-nou tunnel suffered considerable damages in lining concrete such as cracks and flaking and the R/ECC liner construction was applied to particularly damaged sections of total 37 m long. Construction area was 500 m2 and amount of ECC was approximately 25 m^3 (thickness of 50 mm). Works were done night and day on a two-shift system and finished in 10 days (Fig. 3.14).

3.4.3 Use of Tensile Resistance for Steel Floor Deck

Second example using tensile resistance of ECC was a top thickening reinforcement of steel floor deck. Associated with an increase in traffic and heavy vehicles, fatigue damages to steel floor deck have become a serious problem and ECC (see Figs. 3.15 and 3.16) was applied to the top reinforcement [8]. Intended structural merits were increase in stiffness as a hybrid structure of steel deck and ECC and increase in fatigue resistance as a result of stress reduction in steel deck, which was realized by combining a ECC floor slab and existing steel deck with a FRP plate dowel [9]. As an upper thickening material of steel deck, ECC with excellent crack width control capability and tensile resistance capability works satisfactorily if undergone a large tensile strain. In the hybrid floor, thickness of ECC was restricted as thin as 40 mm to avoid a dead load increase and integration with the steel deck must be done within the limited thickness. The FRP plate dowel with a thickness of 10 mm has solved this problem.

This work was realized at the Mihara bridge (Single cable stayed bridge of 972 m long, Ebetsu city, Hokkaido prefecture, Japan) as a fatigue durability upgrading [8]. Completed construction of thickening with ECC is shown in Fig. 3.14. In the

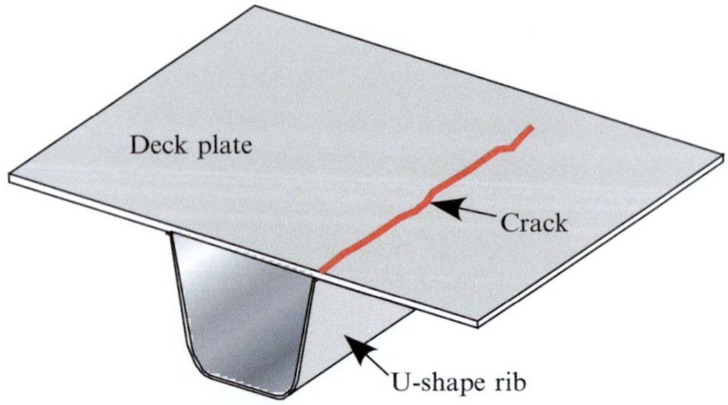

Fig. 3.15 Fatigue damages in steel floor deck

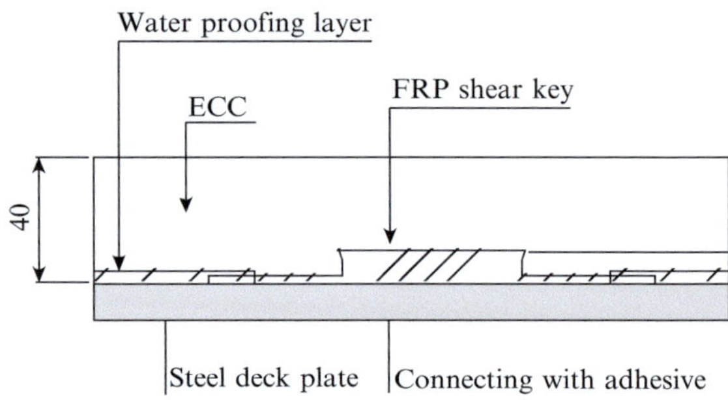

Fig. 3.16 Joint of steel floor deck reinforcing construction

reinforcing construction of the Mihara bridge, a large scale construction was required including mass production of ECC for 800 m^3 and laid down construction of 20,000 m^2. ECC was produced in a ready-mixed concrete plant during exclusive operation at night and the laid down construction was executed with improved concrete paving machines followed by manual troweling. These measures worked well as reported (Fig. 3.17).

3.4.4 Use of Steel Protection Capability for Surface Protection Against Carbonation

For carbonation prevention, paint type lining materials are often applied to reinforced concrete structure for which a long-term durability is required. However under flexural fatigue loads, surface coated members may cause cracks [10] and development of crack width associated with subsequent changes in loads. This results in a problem that the protective capability of the coating would decrease in a very short time within the regular repair interval of e.g. 20 years. ECC could be favorably applied to such cases that surface protective coatings are involved.

ECC protective overlay provides several advantages including control of crack width [11] due for instance to flexural fatigue as shown in Fig. 3.18, prevention of carbonation and decrease in steel corrosion rate.

Fig. 3.17 Thickening construction of bridge floor overlay

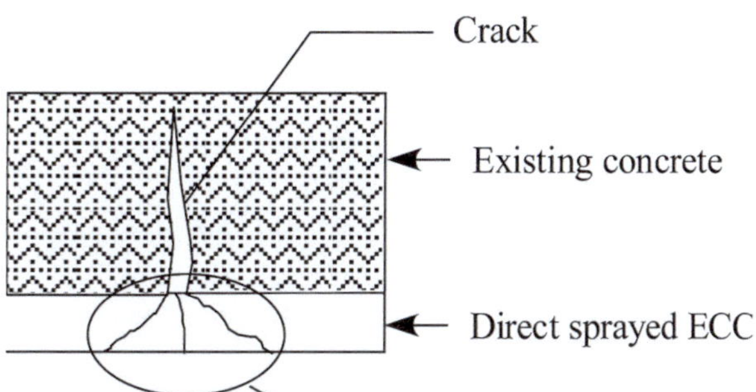

Fig. 3.18 Crack width controlling mechanism of ECC overlay

To verify the crack width control capability of ECC protective overlay, an imitated RC beam was subjected to flexural loading and introduced flexural cracks followed by ECC overlay construction and subsequent repeated loadings. As a result, crack width of ECC protective overlay after repeated loading of 17 million times was just a half of that without ECC and reinforcement as shown in Fig. 3.19. This means that the crack width control capability can also be expected in the case of fatigue loading. Furthermore, the restriction of corrosion rate on the basis of the crack width control capability was confirmed. An example of application of ECC protective overlay is the construction of bottom surfaces of a railway viaduct as shown in Figs. 3.20 and 3.21.

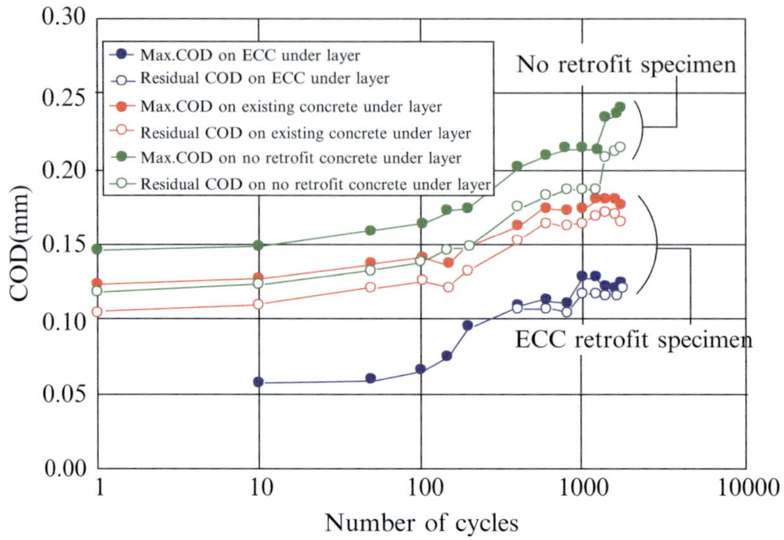

Fig. 3.19 Crack width controlling effects of ECC overlay

Fig. 3.20 Railway viaducts for test construction

Fig. 3.21 Execution of test construction

3.4.5 Seismic Application of Coupled Tensile Resistance and Steel Protection Capability

Application of ECC to seismic members subjected to large deformation has been widely studied [12, 13], for which representing practical applications were documented [14]. In this report, R/ECC joint beams between seismic core walls of high-rise RC building (Fig. 3.22) was targeted, and a design method was developed in compliance with the following two requirements, (1) no substantial strength degradation at a member angle as high as 4% (0.04 rad.) after flexural failure and (2) no cracks influencing durability with a width larger than 0.3 mm after an earthquake. As a result, required performance was attained and two applications were realized. One of them is shown in Fig. 3.23.

3.4.6 Joint and Overlay Applications

Application example other than above, making use of tensile resistance characteristic, is R/ECC PCa hybrid floor, which was designed to meet a requirement for the upgrade of highway bridges. Excellent fatigue durability was confirmed by fatigue test with design wheel loads [15]. Another application of ECC to highway

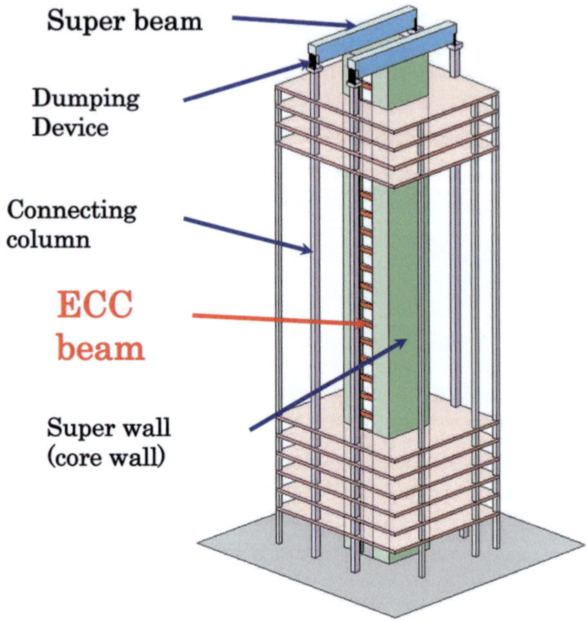

Fig. 3.22 R/ECC joint beams between seismic core walls of high-rise RC building

Fig. 3.23 Realization of R/ECC joint beams

Fig. 3.24 Crack pattern in an ECC repair layer [19]

bridges is a bottom surface thickening of floor slab with particular cases of repair under in-service conditions. A fixed-point loading fatigue test was performed and a possibility of excellent fatigue resistant repair method was suggested [16].

Similar to the application of reinforcing steel protective capability, mass transport reducing capability of ECC overlay has been applied to retaining walls, dams and agricultural channels. Rokugo et al. applied ECC overlay to retaining walls damaged by alkali-aggregate reactions and good results were reported [17]. ECC overlay was also applied to an aged dam whose concrete surface of the upstream side was severely damaged [18]. Some test constructions of ECC applied to agricultural channels were reported.

ECC has also been used for repair layers on cracked concrete surfaces. In coatings on cracked concrete surfaces, crack opening displacements in the substrate may lead to local stress concentrations and cracking within the covering layer. However, repair layers made of ECC appear to be comparably resistant to such loading conditions. Because of their high deformability these materials are capable of bridging cracks in the substrate.

An ECC with a maximum aggregate size of 0.5 mm and a PVA fibre content of 2.2% by volume has been chosen. At first, the crack bridging capacity of this material was investigated in laboratory experiments [19]. Figure 3.24 shows the

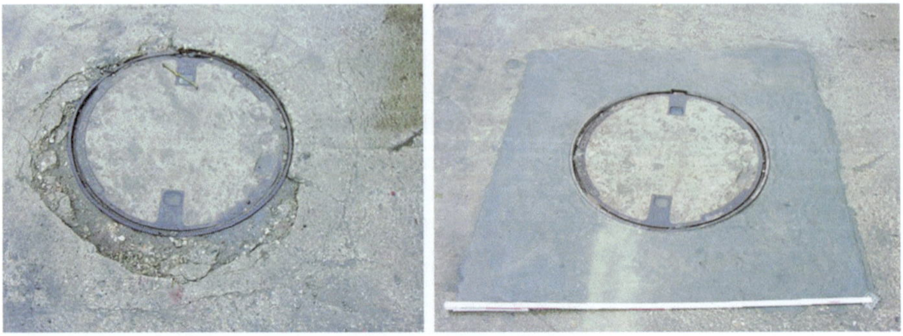

Fig. 3.25 Concrete surface in the vicinity of a sewer before repair (*left*) and 4.5 month after repair (*right*) [19]

crack pattern in a 3 cm thick ECC layer which was applied on the cracked tensile face of an unloaded reinforced concrete beam, see upper part of Fig. 3.24. After the application of the ECC repair layer, the beam was subjected to bending again and it was found that the repair layer could bridge the distinct cracks being opened in the substrate. Under increasing load, a pattern of multiple fine cracks with widths below 50 μm was formed in the ECC layer. Zero-span elongation tests were also performed and yielded the same results. From the laboratory experiments it could be concluded that ECC repair layers with thicknesses of about 3 cm to 4 cm may bridge cracks in a concrete substrate up to a crack opening displacement of about 0.6 mm.

In several pilot applications, the chosen material has been successfully used under site conditions for repair layers on cracked concrete surfaces [19]. An ECC layer with a thickness of about 3 cm was applied on a cracked park deck. Despite noticeable thermally induced crack opening displacements in the concrete substrate, no visible cracks were found on the surface of the repair layer.

Figure 3.25 shows on the left side a concrete slab on grade which was severely damaged in the vicinity of a sewer. Several attempts of patch repair failed before finally ECC was applied as a repair material. The ECC patch, see Fig. 3.25 on the right side, survived five cold seasons so far and could withstand heavy traffic loads.

(T. Kanda, K. Rokugo and V. Slowik)

3.5 Conclusions and Future Tasks

Several actual SHCC application projects have been realized in the world and such application is expected to extend the limit in regular R/C structures. However, application examples introduced in this report were realized owing exclusively to highly skilled engineers and research workers on SHCC who examined materials and structures according to individual design and construction conditions on a case-by-case basis. The present situation does not allow inexperienced engineers to make full use of SHCC due to a lack in technical guidelines. Necessary tasks for SHCC to play an important role in the future are as follows.

1. Determination of standard method for tensile test.
2. Establishment of evaluation method for tensile performance (tensile strength, ultimate tensile train etc.) of materials
3. Evaluation method of crack performance such as crack width and distributions
4. Serviceability limit state and limit state design method taking into account tensile performance of materials (immediate actions for guidelines)
5. Versatile construction and quality control method for large-scale constructions

Application of SHCC is still in incunabula and associated with problems, while it poses a unique feature that has never been presented by the existing cement-based materials. A worldwide active research and development of SHCC imply a possibility to realize concrete structures with excellent safety, serviceability and durability performance.

(T. Kanda and K. Rokugo)

References

1. Fukuyama, H.: Application of high performance fiber reinforced cementitious composites for damage mitigation of building structures – case study on damage mitigation of RC buildings with soft first story. J. Adv. Concr. Technol. **4**, 35–44 (2006)
2. Fukuyama, H., Suwada, H., Mukai, T.: Test on high-performance wall elements with HPFRCC. In: Fischer, G., Li, V.C. (eds.) Proceedings of the International RILEM Workshop on High Performance Fiber Reinforced Cementitious Composites (HPFRCC) in Structural Applications (Rilem Proceedings PRO 49, RILEM Publications S.A.R.L), Honolulu, Hawaii, pp. 365–374 (2005)
3. Fukuyama, H., Iso, M., Ogawa, A., Suwada, H.: Mitigation of damage due to crack of RC elements utilizing high performance fiber reinforced cementitious composites. In: Reinhardt, H.W., Naaman, A.E. (eds.) Proceedings of the Fifth International RILEM Workshop on High Performance Fiber Reinforced Cement Composites (HPFRCCS-5) (RILEM Proceedings PRO53, RILEM Publications S.A.R.L.), Cachan, France, pp. 427–435 (2007)
4. Yamamoto, T., Nagoya, K., Hiraishi, T., Sakata, N., Shimizu, M., Mashimo, H.: Tunnel repair technology using highly ductile cement composites – An application to a tunnel damaged during Chuetsu Earthquake (in Japanese). In: Proceedings of the 60th Annual Meeting (JSCE). J. Mater. Concrete Struct. Pavements **3**, 483–484 (2005)
5. JSCE: Recommendations for Design and Construction of High Performance Fiber Reinforced Cement Composite with Multiple Fine Cracks (HPFRCC). Concrete Engineering Series 82. Japan Society of Civil Engineers, Tokyo (2008)
6. Li, V.C.: From micromechanics to structural engineering – the design of cementitious composites for civil engineering applications. J. Struct. Mech. Earthq. Eng. **10**, 37–48 (1993)
7. Nagoya, K., Mashimo, H., Fukuda, I., Matsubara, K.: Loading test of tunnel repaired using highly ductile cement composites (in Japanese). In: Proceedings of the 60th Annual Meeting (JSCE) J. Mater. Concrete Struct. Pavements **3**, 481–482 (2005)

8. Mitamura, H., Sakata, N., Akashiro, K., Suda, K., Hiraishi, T.: Repair construction of steel deck with highly ductile fiber reinforced cement composites – construction of Mihara bridge (in Japanese). Bridge Found **39**, 88–91 (2005)
9. Fukuda, I., Mitamura, H., Imano, H., Matsui, S.: Top surface thickening with highly ductile cement composites for steel decks with FRP dowels (in Japanese). Proc. Jpn. Concr. Inst. **26**, 1693–1698 (2004)
10. Momiyama, Y., Kimura, H., Yoshida, A.: Durability of coating materials for concrete (in Japanese). Proc. Concr. Struct. Scenar. **1**, 97–102 (2001)
11. Kunieda, M., Inakuma, H., Masukawa, J., Rokugo, K.: Basic study of surface overlay with ECC (in Japanese). Proc. Concr. Struct. Scenar. **4**, 265–270 (2004)
12. Fukuyama, H., Suwada, H.: Experimental response of HPFRCC damper for structural control. J. Adv. Concr. Technol. **1**, 317–326 (2003)
13. Nagai, S., Kaneko, T., Kanda, T., Maruta, M.: Development of seismic control dampers using highly ductile cement composites. KaTRI Annu. Rep. **51**, 57–62 (2003) (in Japanese)
14. Maruta, M., Kanda, T., Nagai, S., Yamamoto, Y.: Realization of new high-rise RC structural frame with precast ECC joint beams (in Japanese). Concr. J. **43**, 18–26 (2005)
15. Suda, K., Kanda, T., Ichimiya, T., Sakata, N., Kaneko, M.: Wheel loading fatigue of highly ductile half-PCa FRC hybrid floor deck (in Japanese). Proc. Concr. Struct. Scenar. **1**, 261–268 (2001)
16. Yokoyama, K., Shikano, Y., Shitou, K., Momiyama, Y.: Evaluation of RC beam repaired with bottom thickening method by fixed point loading fatigue test (in Japanese). Proc. Concr. Struct. Scenar. **2**, 57–60 (2002)
17. Rokugo, K. et al.: Test construction of ECC overlay for gravity type retaining wall and evaluation of tensile performance of elements (in Japanese). In: Proceeding of the Symposium on Highly Ductile Cement Composites, Tokyo, pp. 133–140 (2003)
18. Kojima, S., Sakata, N., Kanda, T., Hiraishi, T.: Application of sprayed repair with highly ductile cement composites to upstream concrete of Sanko dam (in Japanese). Concr. J. **42**, 135–139 (2004)
19. Wagner, C., Slowik, V., Waldenburger, K.: Strain hardening cement-based material for the repair of cracked concrete surfaces (in German). Bautechnik **85**, 49–56 (2008)

RILEM Publications

The following list is presenting our global offer, sorted by series.

RILEM PROCEEDINGS

PRO 1: Durability of High Performance Concrete (1994) 266 pp., ISBN: 2-91214-303-9; e-ISBN: 2-35158-012-5; *Ed. H. Sommer*

PRO 2: Chloride Penetration into Concrete (1995) 496 pp., ISBN: 2-912143-00-4; e-ISBN: 2-912143-45-4; *Eds. L.-O. Nilsson and J.-P. Ollivier*

PRO 3: Evaluation and Strengthening of Existing Masonry Structures (1995) 234 pp., ISBN: 2-912143-02-0; e-ISBN: 2-351580-14-1; *Eds. L. Binda and C. Modena*

PRO 4: Concrete: From Material to Structure (1996) 360 pp., ISBN: 2-912143-04-7; e-ISBN: 2-35158-020-6; *Eds. J.-P. Bournazel and Y. Malier*

PRO 5: The Role of Admixtures in High Performance Concrete (1999) 520 pp., ISBN: 2-912143-05-5; e-ISBN: 2-35158-021-4; *Eds. J. G. Cabrera and R. Rivera-Villarreal*

PRO 6: High Performance Fiber Reinforced Cement Composites (HPFRCC 3) (1999) 686 pp., ISBN: 2-912143-06-3; e-ISBN: 2-35158-022-2; *Eds. H. W. Reinhardt and A. E. Naaman*

PRO 7: 1st International RILEM Symposium on Self-Compacting Concrete (1999) 804 pp., ISBN: 2-912143-09-8; e-ISBN: 2-912143-72-1; *Eds. Å. Skarendahl and Ö. Petersson*

PRO 8: International RILEM Symposium on Timber Engineering (1999) 860 pp., ISBN: 2-912143-10-1; e-ISBN: 2-35158-023-0; *Ed. L. Boström*

PRO 9: 2nd International RILEM Symposium on Adhesion between Polymers and Concrete ISAP '99 (1999) 600 pp., ISBN: 2-912143-11-X; e-ISBN: 2-35158-024-9; *Eds. Y. Ohama and M. Puterman*

K. Rokugo and T. Kanda (eds.), *Strain Hardening Cement Composites: Structural Design and Performance*, RILEM State-of-the-Art Reports 6, DOI 10.1007/978-94-007-4836-1, © RILEM 2013

PRO 10: 3rd International RILEM Symposium on Durability of Building and Construction Sealants (2000) 360 pp., ISBN: 2-912143-13-6; e-ISBN: 2-351580-25-7; *Eds. A. T. Wolf*

PRO 11: 4th International RILEM Conference on Reflective Cracking in Pavements (2000) 549 pp., ISBN: 2-912143-14-4; e-ISBN: 2-35158-026-5; *Eds. A. O. Abd El Halim, D. A. Taylor and El H. H. Mohamed*

PRO 12: International RILEM Workshop on Historic Mortars: Characteristics and Tests (1999) 460 pp., ISBN: 2-912143-15-2; e-ISBN: 2-351580-27-3; *Eds. P. Bartos, C. Groot and J. J. Hughes*

PRO 13: 2nd International RILEM Symposium on Hydration and Setting (1997) 438 pp., ISBN: 2-912143-16-0; e-ISBN: 2-35158-028-1; *Ed. A. Nonat*

PRO 14: Integrated Life-Cycle Design of Materials and Structures (ILCDES 2000) (2000) 550 pp., ISBN: 951-758-408-3; e-ISBN: 2-351580-29-X, ISSN: 0356-9403; *Ed. S. Sarja*

PRO 15: Fifth RILEM Symposium on Fibre-Reinforced Concretes (FRC) – BEFIB'2000 (2000) 810 pp., ISBN: 2-912143-18-7; e-ISBN: 2-912143-73-X; *Eds. P. Rossi and G. Chanvillard*

PRO 16: Life Prediction and Management of Concrete Structures (2000) 242 pp., ISBN: 2-912143-19-5; e-ISBN: 2-351580-30-3; *Ed. D. Naus*

PRO 17: Shrinkage of Concrete – Shrinkage 2000 (2000) 586 pp., ISBN: 2-912143-20-9; e-ISBN: 2-351580-31-1; *Eds. V. Baroghel-Bouny and P.-C. Aïtcin*

PRO 18: Measurement and Interpretation of the On-Site Corrosion Rate (1999) 238 pp., ISBN: 2-912143-21-7; e-ISBN: 2-351580-32-X; *Eds. C. Andrade, C. Alonso, J. Fullea, J. Polimon and J. Rodriguez*

PRO 19: Testing and Modelling the Chloride Ingress into Concrete (2000) 516 pp., ISBN: 2-912143-22-5; e-ISBN: 2-351580-33-8; Soft cover, *Eds. C. Andrade and J. Kropp*

PRO 20: 1st International RILEM Workshop on Microbial Impacts on Building Materials (2000)74 pp., e-ISBN: 2-35158-013-3; *Ed. M. Ribas Silva (CD 02)*

PRO 21: International RILEM Symposium on Connections between Steel and Concrete (2001)1448 pp., ISBN: 2-912143-25-X; e-ISBN: 2-351580-34-6; *Ed. R. Eligehausen*

PRO 22: International RILEM Symposium on Joints in Timber Structures (2001) 672 pp., ISBN: 2-912143-28-4; e-ISBN: 2-351580-35-4; *Eds. S. Aicher and H.-W. Reinhardt*

PRO 23: International RILEM Conference on Early Age Cracking in Cementitious Systems (2003) 398 pp., ISBN: 2-912143-29-2; e-ISBN: 2-351580-36-2; *Eds. K. Kovler and A. Bentur*

PRO 24: 2nd International RILEM Workshop on Frost Resistance of Concrete (2002) 400 pp., ISBN: 2-912143-30-6; e-ISBN: 2-351580-37-0, Hard back; *Eds. M. J. Setzer, R. Auberg and H.-J. Keck*

PRO 25: International RILEM Workshop on Frost Damage in Concrete (1999) 312 pp., ISBN: 2-912143-31-4; e-ISBN: 2-351580-38-9, Soft cover; *Eds. D. J. Janssen, M. J. Setzer and M. B. Snyder*

PRO 26: International RILEM Workshop on On-Site Control and Evaluation of Masonry Structures (2003) 386 pp., ISBN: 2-912143-34-9; e-ISBN: 2-351580-14-1, Soft cover; *Eds. L. Binda and R. C. de Vekey*

PRO 27: International RILEM Symposium on Building Joint Sealants (1988) 240 pp.,e-ISBN: 2-351580-15-X; *Ed. A. T. Wolf (CD03)*

PRO 28: 6th International RILEM Symposium on Performance Testing and Evaluation of Bituminous Materials, PTEBM'03, Zurich, Switzerland (2003) 652 pp., ISBN: 2-912143-35-7; e-ISBN: 2-912143-77-2, Soft cover; *Ed. M. N. Partl (CD06)*

PRO 29: 2nd International RILEM Workshop on Life Prediction and Ageing Management of Concrete Structures, Paris, France (2003) 402 pp., ISBN: 2-912143-36-5; e-ISBN: 2-912143-78-0, Soft cover; *Ed. D. J. Naus*

PRO 30: 4th International RILEM Workshop on High Performance Fiber Reinforced Cement Composites – HPFRCC 4, University of Michigan, Ann Arbor, USA (2003) 562 pp., ISBN: 2-912143-37-3; e-ISBN: 2-912143-79-9, Hard back; *Eds. A. E. Naaman and H. W. Reinhardt*

PRO 31: International RILEM Workshop on Test and Design Methods for Steel Fibre Reinforced Concrete: Background and Experiences (2003) 230 pp., ISBN: 2-912143-38-1; e-ISBN: 2-351580-16-8, Soft cover; *Eds. B. Schnütgen and L. Vandewalle*

PRO 32: International Conference on Advances in Concrete and Structures, 2 volumes (2003) 1592 pp., ISBN (set): 2-912143-41-1; e-ISBN: 2-351580-17-6, Soft cover; *Eds. Ying-shu Yuan, Surendra P. Shah and Heng-lin Lü*

PRO 33: 3rd International Symposium on Self-Compacting Concrete (2003) 1048 pp., ISBN: 2-912143-42-X; e-ISBN: 2-912143-71-3, Soft cover; *Eds. Ó. Wallevik and I. Níelsson*

PRO 34: International RILEM Conference on Microbial Impact on Building Materials (2003) 108 pp., ISBN: 2-912143-43-8; e-ISBN: 2-351580-18-4; *Ed. M. Ribas Silva*

PRO 35: International RILEM TC 186-ISA on Internal Sulfate Attack and Delayed Ettringite Formation (2002) 316 pp., ISBN: 2-912143-44-6; e-ISBN: 2-912143-80-2, Soft cover; *Eds. K. Scrivener and J. Skalny*

PRO 36: International RILEM Symposium on Concrete Science and Engineering – A Tribute to Arnon Bentur (2004) 264 pp., ISBN: 2-912143-46-2; e-ISBN: 2-912143-58-6, Hard back; *Eds. K. Kovler, J. Marchand, S. Mindess and J. Weiss*

PRO 37: 5th International RILEM Conference on Cracking in Pavements – Mitigation, Risk Assessment and Prevention (2004) 740 pp., ISBN: 2-912143-47-0; e-ISBN: 2-912143-76-4, Hard back; *Eds. C. Petit, I. Al-Qadi and A. Millien*

PRO 38: 3rd International RILEM Workshop on Testing and Modelling the Chloride Ingress into Concrete (2002) 462 pp., ISBN: 2-912143-48-9; e-ISBN: 2-912143-57-8, Soft cover; *Eds. C. Andrade and J. Kropp*

PRO 39: 6th International RILEM Symposium on Fibre-Reinforced Concretes (BEFIB 2004), 2 volumes, (2004) 1536 pp., ISBN: 2-912143-51-9 (set); e-ISBN: 2-912143-74-8, Hard back;*Eds. M. Di Prisco, R. Felicetti and G. A. Plizzari*

PRO 40: International RILEM Conference on the Use of Recycled Materials in Buildings and Structures (2004) 1154 pp., ISBN: 2-912143-52-7 (set); e-ISBN: 2-912143-75-6, Soft cover; *Eds. E. Vázquez, Ch. F. Hendriks and G. M. T. Janssen*

PRO 41: RILEM International Symposium on Environment-Conscious Materials and Systems for Sustainable Development (2005) 450 pp., ISBN: 2-912143-55-1; e-ISBN: 2-912143-64-0, Soft cover; *Eds. N. Kashino and Y. Ohama*

PRO 42: SCC'2005 – China: 1st International Symposium on Design, Performance and Use of Self-Consolidating Concrete (2005) 726 pp., ISBN: 2-912143-61-6; e-ISBN: 2-912143-62-4, Hard back; *Eds. Zhiwu Yu, Caijun Shi, Kamal Henri Khayat and Youjun Xie*

PRO 43: International RILEM Workshop on Bonded Concrete Overlays (2004) 114 pp.,e-ISBN: 2-912143-83-7; *Eds. J. L. Granju and J. Silfwerbrand*

PRO 44: 2nd International RILEM Workshop on Microbial Impacts on Building Materials (Brazil 2004) (CD11) 90 pp., e-ISBN: 2-912143-84-5; *Ed. M. Ribas Silva*

PRO 45: 2nd International Symposium on Nanotechnology in Construction, Bilbao, Spain (2005) 414 pp., ISBN: 2-912143-87-X; e-ISBN: 2-912143-88-8, Soft cover; *Eds. Peter J. M. Bartos, Yolanda de Miguel and Antonio Porro*

PRO 46: ConcreteLife'06 – International RILEM-JCI Seminar on Concrete Durability and Service Life Planning: Curing, Crack Control, Performance in Harsh Environments (2006) 526 pp., ISBN: 2-912143-89-6; e-ISBN: 2-912143-90-X, Hard back; *Ed. K. Kovler*

PRO 47: International RILEM Workshop on Performance Based Evaluation and Indicators for Concrete Durability (2007) 385 pp., ISBN: 978-2-912143-95-2; e-ISBN: 978-2-912143-96-9, Soft cover; *Eds. V. Baroghel-Bouny, C. Andrade, R. Torrent and K. Scrivener*

PRO 48: 1st International RILEM Symposium on Advances in Concrete through Science and Engineering (2004) 1616 pp., e-ISBN: 2-912143-92-6; *Eds. J. Weiss, K. Kovler, J. Marchand, and S. Mindess*

PRO 49: International RILEM Workshop on High Performance Fiber Reinforced Cementitious Composites in Structural Applications (2006) 598 pp., ISBN: 2-912143-93-4; e-ISBN: 2-912143-94-2, Soft cover; *Eds. G. Fischer and V.C. Li*

PRO 50: 1st International RILEM Symposium on Textile Reinforced Concrete (2006) 418 pp., ISBN: 2-912143-97-7; e-ISBN: 2-351580-08-7, Soft cover; *Eds. Josef Hegger, Wolfgang Brameshuber and Norbert Will*

PRO 51: 2nd International Symposium on Advances in Concrete through Science and Engineering (2006) 462 pp., ISBN: 2-35158-003-6; e-ISBN: 2-35158-002-8, Hard back; *Eds. J. Marchand, B. Bissonnette, R. Gagné, M. Jolin and F. Paradis*

PRO 52: Volume Changes of Hardening Concrete: Testing and Mitigation (2006) 428 pp., ISBN: 2-35158-004-4; e-ISBN: 2-35158-005-2, Soft cover; *Eds. O. M. Jensen, P. Lura and K. Kovler*

PRO 53: High Performance Fiber Reinforced Cement Composites HPFRCC5 (2007) 542 pp., ISBN: 978-2-35158-046-2; e-ISBN: 978-2-35158-089-9, Hard back; *Eds. H. W. Reinhardt and A. E. Naaman*

PRO 54: 5th International RILEM Symposium on Self-Compacting Concrete, 3 Volumes (2007)1198 pp., ISBN: 978-2-35158-047-9; e-ISBN: 978-2-35158-088-2, Soft cover; *Eds. G. De Schutter and V. Boel*

PRO 55: International RILEM Symposium Photocatalysis, Environment and Construction Materials (2007) 350 pp., ISBN: 978-2-35158-056-1; e-ISBN: 978-2-35158-057-8, Soft cover; *Eds. P. Baglioni and L. Cassar*

PRO 56: International RILEM Workshop on Integral Service Life Modelling of Concrete Structures (2007) 458 pp., ISBN 978-2-35158-058-5; e-ISBN: 978-2-35158-090-5, Hard back; *Eds.R. M. Ferreira, J. Gulikers and C. Andrade*

PRO 57: RILEM Workshop on Performance of cement-based materials in aggressive aqueous environments (2008) 132 pp., e-ISBN: 978-2-35158-059-2; *Ed. N. De Belie*

PRO 58: International RILEM Symposium on Concrete Modelling CONMOD'08 (2008) 847 pp., ISBN: 978-2-35158-060-8; e-ISBN: 978-2-35158-076-9, Soft cover; *Eds. E. Schlangen and G. De Schutter*

PRO 59: International RILEM Conference on On Site Assessment of Concrete, Masonry and Timber Structures SACoMaTiS 2008, 2 volumes (2008) 1232 pp., ISBN: 978-2-35158-061-5 (set); e-ISBN: 978-2-35158-075-2, Hard back; *Eds. L. Binda, M. di Prisco and R. Felicetti*

PRO 60: Seventh RILEM International Symposium (BEFIB 2008) on Fibre Reinforced Concrete: Design and Applications (2008) 1181 pp, ISBN: 978-2-35158-064-6; e-ISBN: 978-2-35158-086-8, Hard back; *Ed. R. Gettu*

PRO 61: 1st International Conference on Microstructure Related Durability of Cementitious Composites (Nanjing), 2 volumes, (2008) 1524 pp., ISBN: 978-2-35158-065-3; e-ISBN: 978-2-35158-084-4; *Eds. W. Sun, K. van Breugel, C. Miao, G. Ye and H. Chen*

PRO 62: NSF/ RILEM Workshop: In-situ Evaluation of Historic Wood and Masonry Structures (2008) 130 pp., e-ISBN: 978-2-35158-068-4; *Eds. B. Kasal, R. Anthony and M. Drdácký*

PRO 63: Concrete in Aggressive Aqueous Environments: Performance, Testing and Modelling, 2 volumes, (2009) 631 pp., ISBN: 978-2-35158-071-4; e-ISBN: 978-2-35158-082-0, Soft cover;*Eds. M. G. Alexander and A. Bertron*

PRO 64: Long Term Performance of Cementitious Barriers and Reinforced Concrete in Nuclear Power Plants and Waste Management – NUCPERF 2009 (2009) 359 pp., ISBN: 978-2-35158-072-1; e-ISBN: 978-2-35158-087-5; *Eds. V. L'Hostis, R. Gens, C. Gallé*

PRO 65: Design Performance and Use of Self-consolidating Concrete, SCC'2009, (2009) 913 pp., ISBN: 978-2-35158-073-8; e-ISBN: 978-2-35158-093-6; *Eds. C. Shi, Z. Yu, K. H. Khayat and P. Yan*

PRO 66: Concrete Durability and Service Life Planning, 2nd International RILEM Workshop, ConcreteLife'09, (2009) 626 pp., ISBN: 978-2-35158-074-5; e-ISBN: 978-2-35158-085-1;*Ed. K. Kovler*

PRO 67: Repairs Mortars for Historic Masonry (2009) 397 pp., e-ISBN: 978-2-35158-083-7;*Ed. C. Groot*

PRO 68: Proceedings of the 3rd International RILEM Symposium on 'Rheology of Cement Suspensions such as Fresh Concrete' (2009) 372 pp., ISBN: 978-2-35158-091-2; e-ISBN: 978-2-35158-092-9; *Eds. O. H. Wallevik, S. Kubens and S. Oesterheld*

PRO 69: 3rd International PhD Student Workshop on 'Modelling the Durability of Reinforced Concrete' (2009) 122 pp., ISBN: 978-2-35158-095-0; e-ISBN: 978-2-35158-094-3;*Eds. R. M. Ferreira, J. Gulikers and C. Andrade*

PRO 71: Advances in Civil Engineering Materials, Proceedings of the 'The 50-year Teaching Anniversary of Prof. Sun Wei', (2010) 307 pp., ISBN: 978-2-35158-098-1; e-ISBN: 978-2-35158-099-8; *Eds. C. Miao, G. Ye, and H. Chen*

PRO 74: International RILEM Conference on 'Use of Superabsorsorbent Polymers and Other New Additives in Concrete' (2010) 374 pp., ISBN: 978-2-35158-104-9; e-ISBN: 978-2-35158-105-6;*Eds. O.M. Jensen, M.T. Hasholt, and S. Laustsen*

PRO 75: International Conference on 'Material Science – 2nd ICTRC – Textile Reinforced Concrete – Theme 1' (2010) 436 pp., ISBN: 978-2-35158-106-3; e-ISBN: 978-2-35158-107-0; *Ed. W. Brameshuber*

PRO 76: International Conference on 'Material Science – HetMat – Modelling of Heterogeneous Materials – Theme 2' (2010) 255 pp., ISBN: 978-2-35158-108-7; e-ISBN: 978-2-35158-109-4;*Ed. W. Brameshuber*

PRO 77: International Conference on 'Material Science – AdIPoC – Additions Improving Properties of Concrete – Theme 3' (2010) 459 pp., ISBN: 978-2-35158-110-0; e-ISBN: 978-2-35158-111-7;*Ed. W. Brameshuber*

PRO 78: 2nd Historic Mortars Conference and RILEM TC 203-RHM Final Workshop – HMC2010 (2010) 1416 pp., e-ISBN: 978-2-35158-112-4; *Eds J. Válek, C. Groot, and J. J. Hughes*

PRO 79: International RILEM Conference on Advances in Construction Materials Through Science and Engineering (2011) 213 pp., e-ISBN: 978-2-35158-117-9; *Eds Christopher Leung and K.T. Wan*

PRO 80: 2nd International RILEM Conference on Concrete Spalling due to Fire Exposure (2011) 453 pp., ISBN: 978-2-35158-118-6, e-ISBN: 978-2-35158-119-3; *Eds E.A.B. Koenders and F. Dehn*

PRO 81: 2nd International RILEM Conference on Strain Hardening Cementitious Composites (SHCC2-Rio) (2011) 451 pp., ISBN: 978-2-35158-120-9, e-ISBN: 978-2-35158-121-6; *Eds R.D. Toledo Filho, F.A. Silva, E.A.B. Koenders and E.M.R. Fairbairn*

PRO 82: 2nd International RILEM Conference on Progress of Recycling in the Built Environment (2011) 507 pp., e-ISBN: 978-2-35158-122-3; *Eds V.M. John, E. Vazquez, S.C. Angulo and C. Ulsen*

PRO 85: RILEM-JCI International Workshop on Crack Control of Mass Concrete and Related Issues concerning Early-Age of Concrete Structures – ConCrack 3 (2012) 237 pp., ISBN: 978-2-35158-125-4, e-ISBN: 978-2-35158-126-1; *Eds F. Toutlemonde and J.M. Torrenti*

RILEM REPORTS

Report 19: Considerations for Use in Managing the Aging of Nuclear Power Plant Concrete Structures (1999) 224 pp., ISBN: 2-912143-07-1; e-ISBN: 2-35158-039-7; *Ed. D. J. Naus*

Report 20: Engineering and Transport Properties of the Interfacial Transition Zone in Cementitious Composites (1999) 396 pp., ISBN: 2-912143-08-X; e-ISBN: 2-35158-040-0; *Eds. M. G. Alexander,G. Arliguie, G. Ballivy, A. Bentur and J. Marchand*

Report 21: Durability of Building Sealants (1999) 450 pp., ISBN: 2-912143-12-8; e-ISBN: 2-35158-041-9; *Ed. A. T. Wolf*

Report 22: Sustainable Raw Materials – Construction and Demolition Waste (2000) 202 pp., ISBN: 2-912143-17-9; e-ISBN: 2-35158-042-7; *Eds. C. F. Hendriks and H. S. Pietersen*

Report 23: Self-Compacting Concrete state-of-the-art report (2001) 166 pp., ISBN: 2-912143-23-3; e-ISBN: 2-912143-59-4, Soft cover; *Eds. Å. Skarendahl and Ö. Petersson*

Report 24: Workability and Rheology of Fresh Concrete: Compendium of Tests (2002) 154 pp., ISBN: 2-912143-32-2; e-ISBN: 2-35158-043-5, Soft cover; *Eds. P. J. M. Bartos, M. Sonebi andA. K. Tamimi*

Report 25: Early Age Cracking in Cementitious Systems (2003) 350 pp., ISBN: 2-912143-33-0; e-ISBN: 2-912143-63-2, Soft cover; *Ed. A. Bentur*

Report 26: Towards Sustainable Roofing (Joint Committee CIB/RILEM) (CD 07), (2001) 28 pp.,e-ISBN: 2-912143-65-9; *Eds. Thomas W. Hutchinson and Keith Roberts*

Report 27: Condition Assessment of Roofs (Joint Committee CIB/RILEM) (CD 08), (2003) 12 pp.,e-ISBN: 2-912143-66-7

Report 28: Final report of RILEM TC 167-COM 'Characterisation of Old Mortars with Respect to Their Repair' (2007) 192 pp., ISBN: 978-2-912143-56-3; e-ISBN: 978-2-912143-67-9, Soft cover;*Eds. C. Groot, G. Ashall and J. Hughes*

Report 29: Pavement Performance Prediction and Evaluation (PPPE): Interlaboratory Tests (2005) 194 pp., e-ISBN: 2-912143-68-3; *Eds. M. Partl and H. Piber*

Report 30: Final Report of RILEM TC 198-URM 'Use of Recycled Materials' (2005) 74 pp., ISBN: 2-912143-82-9; e-ISBN: 2-912143-69-1 – Soft cover; *Eds. Ch. F. Hendriks, G. M. T. Janssen and E. Vázquez*

Report 31: Final Report of RILEM TC 185-ATC 'Advanced testing of cement-based materials during setting and hardening' (2005) 362 pp., ISBN: 2-912143-81-0; e-ISBN: 2-912143-70-5 – Soft cover; *Eds. H. W. Reinhardt and C. U. Grosse*

Report 32: Probabilistic Assessment of Existing Structures. A JCSS publication (2001) 176 pp.,ISBN 2-912143-24-1; e-ISBN: 2-912143-60-8 – Hard back; *Ed. D. Diamantidis*

Report 33: State-of-the-Art Report of RILEM Technical Committee TC 184-IFE 'Industrial Floors' (2006) 158 pp., ISBN 2-35158-006-0; e-ISBN: 2-35158-007-9, Soft cover; *Ed. P. Seidler*

Report 34: Report of RILEM Technical Committee TC 147-FMB 'Fracture mechanics applications to anchorage and bond' Tension of Reinforced Concrete Prisms – Round Robin Analysis and Tests on Bond (2001) 248 pp., e-ISBN 2-912143-91-8; *Eds. L. Elfgren and K. Noghabai*

Report 35: Final Report of RILEM Technical Committee TC 188-CSC 'Casting of Self Compacting Concrete' (2006) 40 pp., ISBN 2-35158-001-X; e-ISBN: 2-912143-98-5 – Soft cover;*Eds. Å. Skarendahl and P. Billberg*

Report 36: State-of-the-Art Report of RILEM Technical Committee TC 201-TRC 'Textile Reinforced Concrete' (2006) 292 pp., ISBN 2-912143-99-3; e-ISBN: 2-35158-000-1, Soft cover; *Ed. W. Brameshuber*

Report 37: State-of-the-Art Report of RILEM Technical Committee TC 192-ECM 'Environment-conscious construction materials and systems' (2007) 88 pp., ISBN: 978-2-35158-053-0; e-ISBN: 2-35158-079-0, Soft cover; *Eds. N. Kashino, D. Van Gemert and K. Imamoto*

Report 38: State-of-the-Art Report of RILEM Technical Committee TC 205-DSC 'Durability of Self-Compacting Concrete' (2007) 204 pp., ISBN: 978-2-35158-048-6; e-ISBN: 2-35158-077-6, Soft cover; *Eds. G. De Schutter and K. Audenaert*

Report 39: Final Report of RILEM Technical Committee TC 187-SOC 'Experimental determination of the stress-crack opening curve for concrete in tension' (2007) 54 pp., ISBN 978-2-35158-049-3; e-ISBN: 978-2-35158-078-3, Soft cover; *Ed. J. Planas*

Report 40: State-of-the-Art Report of RILEM Technical Committee TC 189-NEC 'Non-Destructive Evaluation of the Penetrability and Thickness of the Concrete Cover' (2007) 246 pp.,ISBN 978-2-35158-054-7; e-ISBN: 978-2-35158-080-6, Soft cover; *Eds. R. Torrent and L. Fernández Luco*

Report 41: State-of-the-Art Report of RILEM Technical Committee TC 196-ICC 'Internal Curing of Concrete' (2007) 164 pp., ISBN: 978-2-35158-009-7; e-ISBN: 978-2-35158-082-0, Soft cover; *Eds. K. Kovler and O. M. Jensen*

Report 42: 'Acoustic Emission and Related Non-destructive Evaluation Techniques for Crack Detection and Damage Evaluation in Concrete' – Final Report of RILEM Technical Committee212-ACD (2010) 12 pp., e-ISBN: 978-2-35158-100-1; *Ed. M. Ohtsu*

RILEM Publications published by Springer

RILEM BOOKSERIES (Proceedings)

VOL. 1: Design, Production and Placement of Self-Consolidating Concrete (2010) 466 pp., ISBN: 978-90-481-9663-0; e-ISBN: 978-90-481-9664-7, Hardcover; *Eds. K. Khayat and D. Feyes*

VOL. 2: High Performance Fiber Reinforced Cement Composites 6 – HPFRCC6 (2011) 584 pp., ISBN: 978-94-007-2435-8; e-ISBN: 978-94-007-2436-5, Hardcover; *Eds. G.J. Parra-Montesinos, H.W. Reinhardt and A.E. Naaman*

VOL. 3: Advances in Modeling Concrete Service Life – Proceedings of 4th International RILEM PhD Workshop held in Madrid, Spain, November 19, 2010 (2012) 170 pp., ISBN: 978-94-007-2702-1; e-ISBN: 978-94-007-2703-8, Hardcover; *Eds. C. Andrade and Joost Gulikers*

VOL. 5: Joint fib-RILEM Workshop on Modelling of Corroding Concrete Structures (2011) 290 pp., ISBN: 978-94-007-0676-7; e-ISBN: 978-94-007-0677-4, Hardcover; *Eds. C. Andrade and G. Mancini*

For the latest publications in the RILEM Bookseries, please visit
http://www.springer.com/series/8781

RILEM STATE-OF-THE-ART REPORTS

VOL. 1: State-of-the-Art Report of RILEM Technical Committee TC 207-INR 'Non-Destructive Assessment of Concrete Structures: Reliability and Limits of Single and Combined Techniques' (2012) 390 pp., ISBN: 978-94-007-2735-9; e-ISBN: 978-94-007-2736-6, Hardcover; *Ed. D. Breysse*

K. Rokugo and T. Kanda (eds.), *Strain Hardening Cement Composites: Structural Design and Performance*, RILEM State-of-the-Art Reports 6,
DOI 10.1007/978-94-007-4836-1, © RILEM 2013

VOL. 2: State-of-the-Art Report of RILEM Technical Committee TC 225-SAP 'Application of Super Absorbent Polymers (SAP) in Concrete Construction' (2012) 165 pp., ISBN: 978-94-007-2732-8; e-ISBN: 978-94-007-2733-5, Hardcover; *Eds. V. Mechtcherine and H-W. Reinhardt*

VOL. 3: State-of-the-Art Report of RILEM Technical Committee TC 193-RLS 'Bonded Cement-Based Material Overlays for the Repair, the Lining or the Strengthening of Slabs or Pavements' (2011) 198 pp., ISBN: 978-94-007-1238-6; e-ISBN: 978-94-007-1239-3, Hardcover; *Eds. B. Bissonnette, L. Courard, D.W. Fowler and J-L. Granju*

VOL. 4: State-of-the-Art Report prepared by Subcommittee 2 of RILEM Technical Committee TC 208-HFC 'Durability of Strain-Hardening Fibre-Reinforced Cement-Based Composites' (SHCC) (2011) 151 pp., ISBN: 978-94-007-0337-7; e-ISBN: 978-94-007-0338-4, Hardcover; *Eds. G.P.A.G. van Zijl and F.H. Wittmann*

VOL. 5: State-of-the-Art Report of RILEM Technical Committee TC 194-TDP 'Application of Titanium Dioxide Photocatalysis to Construction Materials' (2011) 60 pp., ISBN: 978-94-007-1296-6; e-ISBN: 978-94-007-1297-3, Hardcover; *Eds. Yoshihiko Ohama and Dionys Van Gemert*

VOL. 7: State-of-the-Art Report of RILEM Technical Committee TC 215-AST 'In Situ Assessment of Structural Timber' (2010) 152 pp., ISBN: 978-94-007-0559-3; e-ISBN: 978-94-007-0560-9,Hardcover; *Eds. B. Kasal and T. Tannert*

For the latest publications in the RILEM State-of-the-Art Reports, please visit http://www.springer.com/series/8780

Index

K. Rokugo and T. Kanda (eds.), *Strain Hardening Cement Composites: Structural Design and Performance*, RILEM State-of-the-Art Reports 6, DOI 10.1007/978-94-007-4836-1, © RILEM 2013

MIX
Papier aus verantwortungsvollen Quellen
Paper from responsible sources
FSC® C105338

If you have any concerns about our products,
you can contact us on
ProductSafety@springernature.com

In case Publisher is established outside the EU,
the EU authorized representative is:
Springer Nature Customer Service Center GmbH
Europaplatz 3, 69115 Heidelberg, Germany

Printed by Libri Plureos GmbH
in Hamburg, Germany